Truly Human Enhancement

Basic Bioethics
Arthur Caplan, editor

A complete list of the books in the Basic Bioethics series appears at the back of this book.

Truly Human Enhancement

A Philosophical Defense of Limits

Nicholas Agar

The MIT Press
Cambridge, Massachusetts
London, England

MIT Press books may be purchased at special quantity discounts for business or sales promotional use. For information, please email special_sales@mitpress.mit.edu.

This book was set in Stone by the MIT Press. Printed and bound in the United States of America.

Library of Congress Cataloging-in-Publication Data.

Agar, Nicholas.
Truly human enhancement : a philosophical defense of limits / Nicholas Agar.
   pages   cm.—(Basic bioethics)
Includes bibliographical references and index.
ISBN 978-0-262-02663-5 (hardcover : alk. paper)
1. Medical innovations—Moral and ethical aspects. 2. Biotechnology—Moral and ethical aspects. 3. Genetic engineering—Philosophy. 4. Medical ethics. I. Title.
RA418.5.M4A33   2013
174.2—dc23
2013016659

10 9 8 7 6 5 4 3 2 1

Contents

Series Foreword

I am pleased to present the forty-first book in the Basic Bioethics series. The series makes innovative works in bioethics available to a broad audience and introduces seminal scholarly manuscripts, state-of-the-art reference works, and textbooks. Topics engaged include the philosophy of medicine, advancing genetics and biotechnology, end-of-life care, health and social policy, and the empirical study of biomedical life. Interdisciplinary work is encouraged.

Arthur Caplan

Basic Bioethics Series Editorial Board
Joseph J. Fins
Rosamond Rhodes
Nadia N. Sawicki
Jan Helge Solbakk

Preface

Our humanity marks the point of convergence of increasingly powerful transformative technologies. Some of these technologies will modify human genetic material. Others will attach cybernetic implants and prostheses to human brains and bodies. This book is a philosophical exploration of the moral and prudential limits on the use of these technologies—specifically on their use to enhance human beings. It presents human enhancement as a good thing, but one that it's possible to have too much of.

I endorse moderate enhancement—the improvement of significant attributes and abilities to levels *within or close to* what is currently possible for human beings. I reject radical enhancement—the improvement of significant attributes and abilities to levels that *greatly exceed* what is currently possible for human beings.

Chapter 1 introduces the notion of a transformative change. Transformative changes alter the state of an individual's mental or physical characteristics in a way that warrants a significant change in how that individual evaluates his or her experiences, beliefs, or achievements. A human being who undergoes a transformative change may find that experiences properly viewed as very valuable prior to the change are significantly less valuable after the change. And vice versa. One gets a false impression of the significance of a transformative change by asking how a subject will feel about the change once he or she has undergone it. I use the process of body-snatching in the iconic movie *Invasion of the Body Snatchers* as an example of a transformative change. Body-snatching is a transformation rightly feared by humans even when told that they will be very happy to have undergone it. I present certain enhancements as species of transformative change.

Chapter 2 explores some motives for human enhancement. It presents two ways in which we can assign value to enhancements. What I call the objective ideal assigns prudential value commensurate with the degree to which a given modification objectively enhances a human capacity. As far as the individual who is a candidate for enhancement is concerned, more is always better. I contrast this with the anthropocentric ideal, which allows that some enhancements of greater objective magnitude are more prudentially valuable than enhancements of lesser magnitude but insists that some enhancements of greater magnitude are less valuable than enhancements of lesser magnitude. Such assessments are appropriate for enhancements of our capacities to levels significantly beyond human norms. Both the objective and anthropocentric ideals endorse human enhancement. I argue that the objective and anthropocentric ideals correspond to two different ways to assign value to human capacities. Our capacities' instrumental value corresponds to the objective ideal; our capacities' intrinsic value corresponds to the anthropocentric ideal.

Chapter 3 explores the different verdicts on the radical enhancement of our physical and cognitive abilities recommended by the objective and anthropocentric ideals. I argue that we can resolve the apparent tension between interventions that preserve our capacities' intrinsic value while forgoing increases in instrumental value and interventions that sacrifice intrinsic value in pursuit of increases in instrumental value. We should focus on making nonhuman technologies more instrumentally valuable. Chapter 4 advances philosophical considerations that are, in effect, the inverse of those presented in chapter 3. It focuses not on the feats enabled by radical enhancement, but instead on radical enhancement's consequences for the identities of those who undergo it. I argue that too much enhancement undermines human identities. It makes our survival over time more precarious.

Chapter 5 tackles a significant rationale for radical cognitive enhancement—the improvement of our species' capacity to do science. More intelligent scientists should be better able to invent technologies that improve our lives. In addition, they should be better able to satisfy a deep and enduring curiosity about the universe and our place in it. This scientific curiosity is held by some to be the defining virtue of our species. I distinguish the science done by unenhanced human scientists from the science done by radically cognitively enhanced scientists in terms of the idealizations that they use. Idealization is an indispensable feature of science. One of its purposes is to simplify reality to make it tractable by cognitively limited explainers. There's good reason to expect a systematic difference between the idealizations of radically enhanced scientists and those used by unenhanced scientists. I argue that we are entitled to place a greater value on unenhanced science. Furthermore, forgoing the degrees of cognitive enhancement that would enable us to do radically enhanced science entails no necessary limit on what we can explain about the universe and our place in it.

Chapter 6 switches focus to the very different topic of radical life extension. My primary example here is work of the dissident gerontologist Aubrey de Grey. De Grey aspires to give humans millennial life expectancies. My criticism of this form of human enhancement differs from those advanced in the previous three chapters. I present a moral argument against the experiments required to make radical life extension a reality. I predict immoral experimentation on the poor and disempowered.

The chapters thus far have focused on the prudential and moral dangers of too much enhancement. Chapter 7 presents a case for moderate forms of human enhancement. I argue that attempts to show that there is something inevitably dangerous about moderate enhancements that involve genetic modification or selection fail. Many valid complaints presented as concerns about how humans are enhanced are better understood as directed at the degree of enhancement.

The primary concern of chapters 8 and 9 is that of enhancements of the moral status of human beings. Beings with higher moral status have a greater entitlement to benefits and stronger protections against harm. I find the prospect of beings with a moral status higher than persons frightening. By radically enhancing human cognitive capacities we may inadvertently create beings whose entitlements to benefits and protections against harms are systematically greater than ours. This will be bad news for the unenhanced. Chapter 8 responds to an argument by Allen Buchanan that it is impossible to enhance status beyond personhood. I present an inductive argument for the possible existence of statuses superior to personhood. Chapter 9 presents an argument that it is wrong to risk producing beings with moral status higher than persons. We should look upon moral status enhancement as creating especially morally needy beings. We are subject to no obligation to create them in the first place. We avoid creating their needs by avoiding creating them.

I conclude this book with a chapter that presents a vision of the human future. This vision embraces technological progress. Perhaps Gene Roddenberry had it right in the original series of *Star Trek*. Here, recognizable human beings use fabulous technologies to travel the universe. They view these technologies very differently from the way they view their brains and bodies. The technologies that transport humans across the universe are radically improved. Human brains and bodies are recognized as grounds of valuable experiences and are preserved.

Acknowledgments

I owe thanks to many people who provided intellectual and emotional support during the writing of this book.

Stuart Brock, Felice Marshall, and Cesar Palacios read the entire manuscript and offered very many philosophically valuable suggestions. Edwin Mares and John Matthewson helped enormously with chapter 5, greatly enhancing my understanding of current debates in the philosophy of science. Mark Walker offered especially cogent advocacy of radical human enhancement that forced me to improve many of my arguments for moderation. David Wasserman helped with the argument in chapter 3. Email exchanges with Allen Buchanan and Tom Douglas improved my argument for the possibility of moral status enhancement presented in chapters 8 and 9. I received valuable comments from Simon Keller and Sondra Bacharach. I'm very grateful to Jan Agar—her careful reading of the final manuscript helped me to avoid many typos (and thinkos). I'd like to thank MIT Press's Phil Laughlin, who was a model of editorial efficiency, Judith Feldmann for her excellent copyediting, and an anonymous referee for the Press, whose comments led to numerous improvements of my arguments.

Finally, I'd like to thank my wife Laurianne, and my sons Alexei and Rafael, who somehow managed to put up with me throughout.

Sections of chapter 3 taken from Nicholas Agar. "Sport, simulation, and EPO." In *The Ideal of Nature: Debates about Biotechnology and the Environment*, ed. Gregory E. Kaebnick. Johns Hopkins University Press, 2011.
Sections of chapter 7 taken from Nicholas Agar. "There is a legitimate place for human genetic enhancement." In *Contemporary Debates in Bioethics*, ed. Art Caplan and Robert Arp. Forthcoming from Wiley-Blackwell.

Chapters 8 and 9 largely based on two pieces in the *Journal of Medical Ethics*:

Nicholas Agar. "Why we can't really say what post-persons are." *Journal of Medical Ethics* 38 (2012): 144–145.

Nicholas Agar. "Why it is possible to enhance and why doing so is wrong." *Journal of Medical Ethics* (forthcoming).

1 Radical Human Enhancement as a Transformative Change

This is the age of human enhancement. Students and pilots swallow pills to enhance their powers of concentration while preparing for exams or operating stealth bombers. Cosmetic surgeons enhance people's appearances. Olympians use artificial means to enhance their sporting performances. These are but a few ways by which human beings signal our dissatisfaction with our evolved (or God-given) abilities.

According to some of its advocates, human enhancement is nothing new. We've been enhancing human capacities ever since we first taught algebra to our children and forced them to eat their sprouts. What's new is the intensity and deliberation with which human enhancement is being pursued. Early twenty-first-century humans enhance themselves in many different ways and by many different means. Technicians of enhancement use scalpels, DNA probes, dietary supplements, and experimental educational techniques to improve our appearances, resistance to disease, rate of aging, and intellects.

The seeming ubiquity of human enhancement doesn't make it right. There's a vigorous debate among philosophers and social critics about the moral acceptability of *any* kind of human enhancement. But the fact that enhancement seems to have become an entrenched feature of affluent early twenty-first-century liberal democracies does suggest the need for a moral inquiry that differs from one directed at the category of human enhancement as a whole, an inquiry sensitive to possible moral differences between human enhancements.

The principal focus of this book is the significance of differences in degree of human enhancement. Its guiding idea is that human enhancement is a good thing, but one that it's possible to have too much of.

A philosophical interest in degrees of human enhancement is timely. A variety of technologies promise human enhancements vastly more powerful than those provided by any pill, injection, supplement, or genetic modification available in the early years of the twenty-first century. Advances in genomics are uncovering genes that influence intelligence, longevity, and many other traits that we may want to enhance. Agricultural biotechnologists modify the genes of our crops and livestock to better suit them to our needs. Similar techniques may boost human intelligence and extend human life spans. Would-be enhancers of humans are not restricted to genetic technologies. Some of them look to cybernetic technologies to enhance by fusing machines to human brains and bodies. The profoundly deaf can now be made to hear, with the attachment to their auditory nerves of cochlear implants. The recipients of these devices would be content to have normal human hearing. But there is no reason that future cochlear implants should not endow them with auditory powers far beyond those conferred by normal biological cochlea. The same goes for any function performed by the human brain. Research into brain–computer interfaces may soon yield implants that dramatically increase our powers of memory or analytical reasoning.

These are exciting prospects. But are they truly desirable? I argue that some ways of enhancing our cognitive powers or of extending our life spans are undesirable specifically because they enhance these attributes to too great a degree. This is so even when lesser degrees of cognitive enhancement or life extension are good. We must distinguish between *moderate* and *radical* degrees of enhancement.

Radical enhancement improves significant attributes and abilities to levels that *greatly exceed* what is currently possible for human beings.[1]

Moderate enhancement improves significant attributes and abilities to levels *within or close to* what is currently possible for human beings.

This book endorses some moderate enhancements but rejects radical enhancement. I call the resulting ideal—a rejection of radical enhancement combined with an endorsement of some moderate enhancements—the ideal of *truly human enhancement*.[2] It accepts human needs and interests as guides in selecting among possible ways to make humans better.

The distinction between moderate and radical enhancement admits of vagueness. Consider the enhancement of our cognitive powers. The futurist and inventor Ray Kurzweil envisages adorning human brains with increasingly many, increasingly powerful electronic implants, soon generating an intelligence "about one billion times more powerful than all human intelligence today."[3] This is radical. A mind that powerful has intellectual capabilities far beyond those of any current human. It's possible to think of cognitive enhancements that fall into a region of vagueness. The *Guinness Book of World Records* credits the Korean civil engineer and former child prodigy Kim Ung-Yong with the world's highest IQ—an intimidating 210. Suppose that IQ is a genuine measure of intelligence. Would an education program that boosted students' IQs to 250 moderately or radically enhance intelligence? Should a human being with an IQ of 250, on the assumption that IQ is a genuine measure of human intelligence, be credited with powers of intellect that greatly exceed what is currently possible for human beings? Perhaps the best thing to say is that it falls into a region of vagueness between moderate and radical enhancement. Vagueness is an obstacle to the adjudication of certain cases. But it doesn't invalidate the categories of radical and moderate enhancement. There are hirsute people and bald people in spite of the lack of a precise boundary between the hairy and the hairless. A degree of vagueness need not prevent good understanding about what makes an enhancement radical and what makes an enhancement moderate. Once we know what to say about the clear cases of radical or moderate enhancement, we should have the intellectual tools needed to tackle the vague cases.

How might enhancements of greater degree be less valuable than those of lesser degree? This book offers a variety of responses to the radical enhancement of human capacities. I will offer *moral* criticisms of some radical enhancements—they impose significant, unjustified costs on others. My arguments here tend to support legal bans on technologies or uses of technologies that lead to certain varieties of radical enhancement. Here, I am principally concerned about enhancements that alter our moral entitlements to various benefits and our moral protections against various harms. Modifications that risk altering our basic moral worth should be banned because they expose unenhanced humans to quite significant harm.

I criticize other radical enhancements from the perspective of *prudential rationality*. The proponents of radical enhancement present it as very prudentially valuable. According to them, the radical enhancement of human mental or physical abilities greatly promotes the well-being and interests of those who undergo it. I argue that radical enhancement will disappoint. We value possible experiences within and slightly beyond the normal human experiential range because we humans can properly engage with them. Possible experiences far outside of the normal human range are less meaningful for us because we're less able to engage with them. They are less prudentially valuable than the experiences that they would replace. We place a lesser value on knowledge enabled by radical cognitive enhancement because we view that knowledge as less effectively extending a distinctively human understanding of the universe and our place in it. Such obstacles either do not exist in respect of lesser enhancements or are significantly reduced. These claims stand in need both of an account of what it means to engage with the experiences and knowledge brought by enhancement and a theory about why successes or failures of engagement should be so important.

The claim that certain kinds of radical enhancements are immoral suggests a response that differs from that supported by the claim that it is prudentially irrational to radically enhance. Laws may be required to prevent immoral radical enhancements—those that impose significant, unjustified costs on others. Radical enhancements that are prudentially irrational may not require legal prohibitions. The law permits individuals to perform a variety of imprudent acts, such as always eating only ice cream and investing large sums of money in astrological divinations. So perhaps citizens should be permitted to enhance themselves in ways that are predictably bad for them. But the imprudence of many radical enhancements does suggest the need for public advisories. Laws may be required to protect children whose legal guardians procure radical enhancements for them in the mistaken belief that they will be good for them. Laws may also be required to counter distorted representations of radical enhancement by its purveyors.

Why write about radical enhancement now? One reason is that it may be imminent. According to some commentators, we're on the verge of quite extreme enhancements of human capacities. For example, Ray Kurzweil proposes a law of accelerating returns according to which the advancement of enhancement technologies follows an exponential path.[4] A feature of this ever-increasing pattern of improvement is that it can deliver dramatic improvements quickly. The law seems particularly well suited to advances in information technologies. Kurzweil predicts that improvements of information-processing technologies will lead humans to fuse with machines. Perhaps he's exaggerating. Perhaps the technologies of human enhancement are not becoming more powerful at an exponential rate. Nevertheless, our humanity is at a point of convergence of a wide variety of technologies. Human enhancement has become a guiding focus for many of those developing genetic and cybernetic technologies. We see evidence for this in the science sections of our newspapers that carry increasingly many stories about technological advances that may lead to enhanced humans. It would be a serious mistake to just assume a scenario in which enhancements arrive in small increments with plenty of time for us to adjust between each installment.

Advances in ethical understanding conform to a very different schedule from that kept by technological advances. There is something necessarily time-consuming about the proper moral appraisal of a technological novelty. A skilled mathematician may rapidly and efficiently deploy her mathematical understanding to address a genuinely soluble novel mathematical problem. Ethical expertise differs in drawing on a very wide and diverse collection of sources of information.[5] Questions about the morality or prudential advisability of human enhancement draw on information about how enhancement technologies will work, how they will be tested, the resources required for their manufacture, and their likely cost. This information must be considered in light of the needs and interests of the individuals and collections of individuals who are the subjects of enhancement or are indirectly affected by enhancement. There are no shortcuts in a proper ethical evaluation. Too cursory or casual an evaluation of a key line of evidence can lead to erroneous conclusions. A widely acknowledged tendency for technological progress to outpace ethical understanding reflects differences in the natures of the tasks. Those engaged in improving a technology are able to focus their attention on the specifics of its design, whereas ethical evaluation is necessarily broad. Each new iteration of an enhancement technology potentially changes what it means for affected individuals and collections of individuals. All of this strongly suggests the need to get underway our ethical evaluation of possible future enhancements as soon as possible.

Transformative Change and *Invasion of the Body Snatchers*

This book presents radical enhancement as a *transformative change*:

A transformative change alters the state of an individual's mental or physical characteristics in a way that causes and warrants a significant change in how that individual evaluates a wide range of their own experiences, beliefs, or achievements.

Transformative changes bring new ways to evaluate experiences, beliefs, or achievements. The changes are not isolated or piecemeal. They occur across a wide range of experiences, beliefs, or achievements. Considered individually, many of the changes in evaluation are significant rather than minor. The changes respond specifically to and are warranted by alterations of a subject's mental or physical characteristics. They do not occur as a response to changes of the world external to a subject's mind and body. According to the view presented in this book, what's truly noteworthy about a doubling of our cognitive powers is not so much that this would be a human enhancement. Rather, what's noteworthy is that it is a significant change that warrants a new evaluative perspective on our own experiences, beliefs, or achievements.

To better understand the perils that follow from radical enhancement's status as a transformative change, I begin with discussion of a thought experiment involving a particularly stark case of transformative change that does not involve enhancement. Philosophers' thought experiments often deliberately simplify reality so as to make key facts especially salient. I propose that some of the conclusions we draw about the following science fiction story apply to the more complex case of transformative change by radical enhancement.

The original 1956 version of the movie *Invasion of the Body Snatchers* is set in a small California town that is the focus of an incursion of extraterrestrials altogether more subtle than the shock-and-awe versions depicted in most other exemplars of the alien invasion genre. The 1956 movie was followed by a memorable remake in 1978 and a less memorable one in 1993. The movies depict the aliens originating in pod form. When placed near a sleeping human, the pods begin the process of snatching the human. They destroy the body of the targeted human, producing a near exact physical and psychological duplicate as they do so. The newly minted pod-person sets about turning more humans into pod-people.

What happens to human beings when they are body-snatched? One possibility is the original human individual dies and is replaced by a pod-person imposter. Robert Nozick defends a different view of the process. He arrives at the conclusion that the human individuals who are snatched do survive the process. The fact that they do not survive *as humans* does not prevent them from surviving *as individuals*. The idea that we as individuals might survive the loss of our humanity will not be surprising to some religious believers who hold that we can survive as disembodied souls. Nozick notes that the process that produces a pod-person from a human person does so in a way that permits no overlap—the act of translating a pod into human form shrivels up the human body of which it is a copy. The end of the human stage of your existence coincides with the beginning of the pod-person stage of your existence. There are deep and nonaccidental psychological similarities too. The pod-person psychology is "explained by and duplicates the one already existing except for the difference in affect."[6] Snatching seems to blunt its subjects' emotions, but it does transfer a wide range of psychological states including, most importantly, autobiographical memories. Autobiographical memories are central to our sense of ourselves as persisting through time. The pod-person whom Nozick supposes I might become would remember the events of my childhood in precisely the same first-person way that I currently remember them. He would not be like some imposter who merely pretends to remember my childhood, responding to questions as if he had genuine memories. The new desires, including most importantly the desire to convert other humans into pod-people and the flattening of affect, do not pose insuperable barriers to the transmission of our identities. People survive religious conversions that endow them with the powerful desire to convert others. They also survive taking pharmacological agents that flatten their affective states.

Suppose that body-snatching does preserve the identities of the human beings who undergo it. It seems straightforwardly to satisfy our criteria for a transformative change. Snatching alters its victim's mental and physical characteristics in a way that warrants a significant change in how that victim evaluates a wide range of her own experiences, beliefs, or achievements. The body-snatched humans do not come to believe in the goodness of life as a pod-person as a result of arguments presented to them by other pod-people—though, as we shall see, the pod-people do present arguments. Rather, the change in evaluative belief is warranted by a change in mental or physical characteristics. Snatching warrants altering a wide range of beliefs about what is good for you.

There is spirited philosophical debate over the metaphysics of personal identity. Some accounts of personal identity will treat body-snatching as killing its human subjects.[7] These theories make much of the fact that the pod-person is physically discontinuous with the snatched human. The matter that constitutes the pod-person's brain originates from a plantlike pod object. I will have a good deal more to say about the implications of transformative change for identity. For now I note that there are cases of transformative change in science fiction that do preserve a human subject's brain. For example, the cybermen, a race of cyborg baddies with frequent guest spots in the long-running BBC TV series *Doctor Who*, subject humans to a transformation known as cyberconversion. This process "upgrades" humans by extracting their brains and rehousing them in machine exoskeletons—it turns humans into cybermen. The process is accompanied by screams, giving evidence that the human candidate for cyberconversion remains conscious throughout the procedure. A collection of distinctively human values is replaced by a collection of values oriented toward domination of the galaxy and inducting additional recruits into the cyberarmy. In *Star Trek*, a collective of malevolent beings known as the Borg subject humans to the transformative change of assimilation. This process leaves largely intact human subjects' brains and bodies. Assimilated humans are modified with a variety of implants connecting them to a group Borg mind. They set about expanding their empire by assimilating additional, formerly autonomous beings.

Suppose that we, as individuals, do survive body-snatching, cyberconversion, or Borg assimilation. In each case, we undergo transformative changes. The evaluative frameworks we apply to our lives—and to the rest of the world—are significantly altered. We place very different values on characteristically human experiences, beliefs, or achievements prior to the change than we do after it. This change seems warranted by changes to the states of our minds and bodies.

This book treats radical enhancement as a change to human beings that shares some of the features of body-snatching, cyberconversion, and Borg assimilation. Cyberconversion and Borg assimilation are, in addition to being transformative changes, both cases of radical enhancement. Physical and cognitive enhancement and life extension are the original motives for developing the technology of cyberconversion. Borg assimilation makes its subjects part of a very powerful group mind.

There is one distracting difference between the transformative changes of body-snatching, cyberconversion, or Borg assimilation, on the one hand, and the manner of radical enhancement sought by humans, on the other hand. The pod-aliens, cybermen, and Borg are science fiction baddies with evil agendas. Body-snatching may be bad for you. But it's also very bad for other human beings. Your snatching increases the likelihood that they will experience the same misfortune. The primary motive of the cybermen is not to bring the benefits of life extension and enhanced cognition to human candidates for cyberconversion but instead to gain recruits for an army of intergalactic domination. The same goes for the Borg. People who seek radical enhancement for themselves or for others need have no such agendas.

We can ask two different kinds of question about the goodness of a particular transformative change. Our questions could address possible conflicts between the change and our moral values. Body-snatching, cyberconversion, and Borg assimilation straightforwardly conflict with our moral values. They increase the likelihood that bad things will happen to other people. Some commentators think that enhancement may have morally disastrous consequences.[8] But many of the questions asked in this book address a different sense in which transformative changes could be good or bad: transformative changes might be prudentially good or bad. Questions about prudential value specifically address consequences for the subjects of a particular transformative change. When asking about the prudential rationality of body-snatching, cyberconversion, and Borg assimilation, we pointedly exclude effects on others. We are interested in whether these transformations are good for those who undergo them.

Body-snatching, cyberconversion, and Borg assimilation warrant significant changes in how an individual evaluates a wide range of his experiences, beliefs, or achievements. Experiences that were formerly good—the feeling of love, for example—become bad. Achievements of which one was formerly proud, such as eluding body-snatching for so long, become, at best, pointless. Radical enhancement manifests these same features, albeit less starkly. What seems creepy about body-snatching, cyberconversion, and Borg assimilation is the subjection of human bodies and brains by alien psychologies. Here an invading psychology is alien in the sense that it is extraterrestrial. There is no extraterrestrial source for the goals, pleasures, and values brought by radical enhancement. But they are alien in the sense that they should be viewed as having a provenance from outside of the individual who acquires them. Their source is not invading extraterrestrials but rather the technologies of enhancement.

The Rational Irreversibility of Some Transformative Changes

At various points in both the 1956 and 1978 *Invasion of the Body Snatchers* movies, pod-people attempt to convince humans that snatching is in their interests. In the 1956 movie, the pod-person Dr. Dan "Danny" Kauffman assures potential inductees that snatching does not really take much of value from them. He says, "Love, desire, ambition, faith—without them, life's so simple, believe me." Freed from these often vexing emotions, the pod-people live harmoniously, working toward common goals. Dr. Kauffman presents the shedding of these emotions as good for the subjects of body-snatching. According to him, it would be prudent to consent to the process. It's reasonable to think that Dr. Kauffman speaks sincerely when he says that he finds his life better without love, desire, ambition, and faith. He is not engaging in deliberate deception.

We might interpret Dr. Kauffman as seeking to appeal to objective truths about the relative value of different kinds of lives. Such truths would enable principled adjudications of the question about which of the two lives is better—the human life that Dr. Kauffman left behind or the pod-person existence he now enjoys. Once we have access to these objective standards we might pronounce a life with love, desire, ambition, and faith better or worse than one without them.

The possibility of appealing to objective standards capable of telling us which transformative changes are good for us and which aren't seems to be supported by a famous remark of John Stuart Mill: "It is better to be a human being dissatisfied than a pig satisfied." He allowed that the pig may disagree. But this disagreement counts for less than the assessment of the dissatisfied human being because it knows only its "own side of the question. The other party to the comparison knows both sides."[9]

Mill certainly didn't take himself to be giving advice to people contemplating transformative change either for themselves or for pigs. His concerns were theoretical rather than practical. But Mill's conclusion should apply to the results of a transformation. If there are facts of the matter about which of two different kinds of existence is better, then it could be prudentially rational to pursue certain transformative changes. Suppose that it were possible for the pig to be turned into a human. Mill would say that it would be in the pig's interests to undergo the transformation. It's plausible that the pig could survive such a transformation if it occurred by way of a series of genetic alterations and neurosurgeries that preserved the pig as a subject of consciousness. The alterations and surgeries might even preserve some autobiographical memories—albeit in a somewhat scrambled state. Of course, there's no possibility that the pig could consent to the procedure or even understand it. But those who own pets subject them to a variety of procedures that we evaluate to be in their interests even when those particular interests are beyond the pet's understanding. We presume that medical treatments for pets are often in their interests even when the pets have no understanding of them and therefore cannot consent.

One premise of Mill's argument seems suspect. Mill supposes that greater cognitive complexity of humans gives them insight into the pig's less complex psychology. I wonder whether dissatisfied humans can really be said to know the pig's "side of the question" or whether Mill's crediting the human with this knowledge assumes that because people are capable of acting in ways that we call piggish—gorging on pizza, or forgoing baths—we arrogantly credit them with insight into what it is like to be a pig.

I propose that what may be the case for pigs is not true for beings who exceed a threshold of cognitive complexity. Individuals who are capable of making rational choices about the direction that their lives will take should have these life plans treated as authoritative. Consider two categories of being, *A* and *B*. The members of *A* are similar in some respects but dissimilar in others. The members of *A* and *B* exceed a threshold of cognitive complexity required for a life plan, a view about which kind of things make their lives go better or worse. The members of both groups satisfy the criteria for personhood. They are self-conscious, rational, and aware of themselves as persisting through time. They have preferences about their own futures. But *A* and *B* are very different kinds of persons. They have different views about the experiences, beliefs, and activities that comprise a good life. Further, suppose that the members of *A* differ from the members of *B* in having superior cognitive abilities, longer life expectancies, and so on. Whether it would be good for a member of *B* to undergo a transformative change turning it into a member of *A* cannot be decided by these facts alone. Claims about the goodness of the transformative change must appeal to the life plans or rational choices of the particular member of *B*. The individual who is a candidate for a transformative change from *B* to *A* must genuinely value the respects in which members of *A* are superior. This valuing may not be explicit, but it must be, at a minimum, implicit, able to be inferred from acknowledged values. If the considered choices of individuals belonging to B did not place importance on the respects in which members of A are superior, then the transformative change should not be viewed as good for it.

If Dr. Kauffman had any interest in doing right by human candidates for body-snatching he should ask whether this transformative change was implied by or is consistent with the rational desires and life plans of the humans. To answer this question in the negative is not to say that there is something objectively wrong about being a pod-person. It is to assert that pod-people are bad kinds of beings for humans to become.

Note, however, that the fact that humans universally and unhesitatingly reject body-snatching does not necessarily mean that we are right to do so. Perhaps we are mistaken about what we value. Perhaps we are victims of status quo bias—we stick with the human condition even when it should be clear to us that body-snatching would bring genuine improvements.[10] We should not suppose that the evaluative framework implied by a life plan is immune to error. Some life plans are founded on factual errors. Suppose you commit to a life plan involving the consumption of large quantities of red wine in the belief that it will extend your life. In fact, in the quantities that you consume it, the wine shortens your life. Your commitment to a long and healthy life rationally requires you to alter your attitudes toward alcohol. There does not seem to be any readily identifiable mistake of this sort in the human rejection of body-snatching. It's hard to imagine Dr. Kauffman successfully supplying truths about what it's like to be a pod-person that should convince humans to submit to body-snatching. Facts about future desires are irrelevant if they derive from the evaluative framework brought by the transformative change. In the 1978 version of the movie, Dr. Kibner, a psychoanalyst played by Leonard Nimoy, inherits from Dr. Kauffman the role of body-snatching's chief proselytizer. He is about to arrange the snatching of Elizabeth Driscoll. Kibner responds to her challenge that humans will stop the pod-aliens with the observation that "in an hour, you won't want them to. In an hour, you'll be one of us." This claim about Elizabeth's future desires and values is surely true—supposing that she actually is snatched. But it should not pretend to inform her about what she values or should desire now. It draws support from an evaluative framework that differs from her current, human one.

The rejection of any independent standard to adjudicate the relative merits of the lives of different varieties of persons means that certain experiences and ways of being properly valued by members of one variety of person may not be valuable to members of other varieties of person. There is no independent framework from which to justify to humans and pod-aliens any single collection of experiences, beliefs, and achievements as the best for all.

Suppose we are given a description of the state of a being at a given time. We might assess that state as pleasant, unpleasant, or somewhere in between. But that assessment falls short of telling us whether what we have described is really good for certain kinds of rational being. Some objectively acceptable ways to be are good for certain kinds of being to become or instantiate. They are bad for other kinds of being to become or instantiate. We should allow that some experiences and ways of being properly valued by humans are not properly valued by pod-aliens. These include love, desire, ambition, and faith. There is no argument that should force Dr. Kauffman to recognize the great value of these human emotions and experiences. For him, they will be icky distractions from worthwhile undertakings. Humans should take the same attitude toward many experiences and ways of being properly valued by pod-aliens. We should assess a given transformation relative to the interests of beings who are candidates for it. It gives weight to how subjects might feel about a transformative change only in respects that their possible future experiences matter to them now.

Consider the possibility of reversing the transformation. If the points in the previous paragraphs are true, then restoring Dr. Kauffman to a human existence could be a very bad thing for him. This is so even if, once he has undergone it, he would be exceedingly grateful. For the reverse transformation to be a prudentially good thing for Dr. Kauffman, this goodness would have to find expression, either explicit or implicit, in his life plan. Dr. Kauffman is no longer human, but he does seem to be a rational being able to form a view about what kinds of things are good for him. He is morally blameless in respect of his own transformative change. Turning him back into a human being would restore to him a whole collection of emotions that he rightly feels happy to be rid of.

We might morally justify reversing body-snatching to prevent the aliens from imposing pod-existences on us. But we should not pretend that this act is good for the pod-person who undergoes it. Suppose the pod-aliens had snatched a hundred humans and were content to go no further. A pod-alien is the *result* of an immoral act perpetrated on a human. It is not the *cause* of that act. We are supposing that the pod-aliens pose no further threat to their human fellow citizens. The beings these people have become no longer value the kinds of experiences that they formerly valued. Reversing the procedure could be a bad thing for those hundred even if it resulted in entirely content human individuals.

One especially chilling feature of transformative changes becomes apparent. Many of them are *rationally irreversible*. Suppose you mistakenly undertake a nontransformative change. You submit yourself to a cosmetic procedure that removes every hair on your body. You can regret your choice and undertake to reverse it. In this case you may have to resort to wearing a wig while you await the regrowth of your hair. But the change is straightforwardly rationally reversible. A transformative change that brings an evaluative scheme that endorses the change necessarily prevents you from acknowledging your error. The change may be practically reversible—others may undertake to change you back. They may succeed. But you are prevented from regretting your choice. The existence of good transformative changes means that you should not reject a change solely because it is transformative. But the rational irreversibility of some of these changes makes it incumbent upon you to choose wisely.

Positive and Negative Transformative Changes

Body-snatching is an example of a transformative change that conflicts with the values of human beings who are candidates for it. I have argued that it is rational for them to oppose it even if body-snatching would turn them into beings who were very glad to have undergone the process. The transformed state is incompatible with the values of the candidate for transformation.

A human fear of body-snatching makes sense given what we value about our lives. Proper understanding of our values may lead us to take a positive view of other transformative changes.

A caterpillar turns into a chrysalis and thence into a butterfly. These are not transformative changes according to the definition advanced in this book. None of caterpillars, chrysalises, nor butterflies are capable of entertaining views about the goodness or badness of their lives or these changes.

But, suppose that a caterpillar were capable of evaluating these changes. We can imagine that it might endorse them. The caterpillar-stage would understand that the butterfly-stage would hold different views about the value of its experiences, beliefs, or achievements. The butterfly-stage might predictably disdain some aspects of being a caterpillar—the pleasures of lounging on luscious green leaves. But the rational caterpillar might nevertheless desire to undergo this change. It would view much of what it did as oriented toward achieving this particular transformation.

A more straightforward case of transformative change occurs as human children mature. Many quite dramatic changes occur as human babies become human adults. Human babies are incapable of entertaining views about the value of their experiences, beliefs, or achievements. But older children are. Their views about the value of their experiences, beliefs, or achievements change quite significantly as they mature. Maturing toward adulthood is, nevertheless, a positive transformative change for most children to undergo. Children actively desire to become the kinds of beings that their parents are. They are generally not deterred by the realization that they may lose their fascination for *Teenage Mutant Ninja Turtles*. Many desires that make no explicit reference to adulthood nevertheless imply a desire to become an adult. Fire hoses and stethoscopes are not easily operable by children, and this means that children who desire to become firefighters or doctors do implicitly desire to grow up. I present radical enhancement as different from this case. It conflicts with rather than promotes our deepest self-regarding desires. In chapter 4, I will have more to say about an attempted philosophical comparison of radical enhancement with growing up.

Radical Enhancement as a Negative Transformative Change

In this book I present radical enhancement as a negative transformative change—it changes for the worse the way that we evaluate the experiences, beliefs, and achievements that constitute our lives.

The first task is to show that radical enhancement really does bring with it a new evaluative framework. This is certainly not the view of Ray Kurzweil. Kurzweil presents radical enhancement as preserving our principal aesthetic and moral values. For example, he says that even "with our mostly nonbiological brains we're likely to keep the aesthetics and emotional import of human bodies, given the influence this aesthetic has on the human brain."[11] If Kurzweil is right, then radical enhancement may not count as transformative change, at least not in the way that I have defined the concept. Kurzweil signals a continuity by endorsing the word "human" as a description for the machine-beings he thinks we will become. The following chapters show that the act of radically enhancing will, almost certainly, alter our evaluative frameworks. We will take a different attitude to the experiences and achievements that we formerly valued.

The second task is to show that radical enhancement is a negative transformative change. Demonstrating that radical enhancement is a negative transformative change does not require that the radically enhanced state is bad in some objective sense. It's possible that a transformed state is good without being good for a human being to become. The state of being a contented pod-person may be a good way to be without being good for a human to become. The state of being a contented chimpanzee is good. But this does not imply that it is good for a human to become a contented chimpanzee. A radically enhanced state may be very good. Furthermore, it may be very good for certain kinds of beings to achieve. I shall argue that, appearances notwithstanding, it is not good for humans to achieve.

There seems to be a strong *prima facie* case for viewing radical enhancement as a positive rather than a negative transformative change. Currently, many people express the desire to radically enhance themselves. In the following chapters I show that radical enhancement is a negative transformative change by subjecting to scrutiny human desires for enhancement. The complexity of financial markets leads people to make investment choices that do not correspond with their desires. I argue that, similarly, complexities in the value that we place on our lives can lead to a mistaken understanding of radical enhancement. Proper information about our values leads us to renounce a desire for radical enhancement.

2   Two Ideals of Human Enhancement

This chapter presents two ideals that compete to direct the enhancement of human beings. According to the *objective ideal*, an enhancement has prudential value commensurate with the degree to which it objectively enhances a human capacity. Technologies that produce enhancements of greater objective magnitude are, all else equal, more valuable than technologies that produce enhancements of lesser magnitude. The objective ideal is strongly suggested by many of the statements of the members of an intellectual and cultural movement known as transhumanism. According to the *anthropocentric ideal*, some enhancements of greater objective magnitude are more prudentially valuable than enhancements of lesser magnitude. However, some enhancements of greater magnitude are less valuable than enhancements of lesser magnitude. Such assessments are warranted for enhancements of our capacities to levels significantly beyond human norms. Both the objective and anthropocentric ideals endorse human enhancement. They are therefore properly contrasted with the bioconservative view that purports to reject all forms of human enhancement.[1] I respond to the blanket bioconservative rejection of enhancement in chapter 7.[2]

The philosophical task would be straightforward if we had merely to decide which of the objective or anthropocentric ideals was correct. We would then insist that, insofar as they are seeking to enhance humans, designers of enhancement technologies conform to the dictates of this true ideal of human enhancement. That will not be my conclusion. Rather, I argue that the objective and anthropocentric ideals correspond to two legitimate ways of assigning prudential value to human capacities—they imply two different ways in which the enhancement of our capacities can promote our interests or well-being.

We can view our capacities as *instrumentally valuable*. Our cognitive capacities enable us to solve difficult problems. We use our muscles to lift heavy objects. The objective ideal describes the effect of enhancement technologies on our capacities' instrumental value. Cognitive enhancements of greater objective magnitude enable more difficult problems to be solved, and objectively greater enhancements of physical strength enable heavier objects to be lifted. Enhancements therefore increase our capacities' instrumental value.

We can also view our capacities as *intrinsically valuable*. This value is independent of the results we use our capacities to bring about—it corresponds with an engagement that we feel with exercises of our capacities. I use some observations of Alasdair MacIntyre about the internal goods of our activities to clarify this somewhat vague-sounding statement.

This chapter and chapters 3, 4, and 5 focus on the enhancement of our cognitive and physical capacities. These forms of enhancement have as their principal motivation new kinds of experience or achievement. They differ from the radical life extension that will have been achieved if we have much longer lives to fill with the kinds of human experience and achievement that are currently available to us. Radical life extension is the topic of chapter 6.

Defining Human Enhancement

An account of radical human enhancement stands in need of an account of human enhancement. It's possible to identify two basic accounts of what it means to enhance a human being in the philosophical literature. The broadest concept of human enhancement identifies it with improvement.[3] To enhance a human being is to improve him or her. The genetic enhancement of intelligence improves its subject's intelligence by means of modifying genes.

This approach requires a principled account of human improvement. What one person may view as an improvement another may view as a worsening. But even if we have such an account, a definition of enhancement as improvement seems too broad to highlight many of the philosophical issues raised by human enhancement. Once defined as improvement, human enhancement becomes unobjectionable. Enhancement as improvement seems to be an indispensable part of being human. We attend universities in the hope that they will improve our minds. We seek to improve our health by consuming omega 3 tablets. Enhancement as improvement is not something that we can realistically contemplate rejecting. As Nick Bostrom and Julian Savulescu point out, "stripped of all such 'enhancements' it would be impossible for us to survive, and maybe we would not even be fully human in the few short days before we perished."[4]

A definition that identifies human enhancement with human improvement risks shortchanging opponents of human enhancement. It seems to require opponents of human enhancement to reject teaching algebra to children, for example. One way to minimize the risk of mishearing the opponents of human enhancement is to allow that their complaints may be better expressed by means of a different concept of enhancement. We would, in effect, endorse a conceptual pluralism that acknowledges the need for more than one concept of human enhancement to address the hugely varied ways in which humans may change. We would recognize *enhancement as improvement*, but acknowledge the need for other concepts of human enhancement perhaps better suited to expressing certain legitimate concerns about how technology may reshape human beings.

A concept better suited to expressing the concerns of opponents of human enhancement appeals to human norms. I propose calling it *enhancement beyond human norms*. According to this account, the modification of a human capacity counts as an enhancement only if it enhances beyond human norms. The norms in question are biological.[5] This concept contrasts enhancement with therapy, which includes measures designed to restore or preserve normal levels of biological functioning. Doctors prescribe synthetic erythropoietin (EPO) to counter the anemia resulting from chronic kidney disease. This is a therapeutic use of EPO—it has the purpose of restoring a patient's red blood cells to biologically normal levels. The therapeutic use of EPO satisfies the requirements of enhancement as improvement. Doctors are seeking to improve their patients' health. Enhancement beyond human norms encompasses interventions whose purpose is to boost levels of functioning beyond biological norms. EPO can serve therapeutic purposes. But it can also be put to nontherapeutic ends. EPO grants competitive advantages to Tour de France cyclists by boosting their levels of red blood cells. Their higher-than-normal levels of red blood cells give them superior endurance—especially useful for the infamously grueling hill stages of the Tour. For them, EPO is a means of enhancement beyond human norms.

The preceding paragraphs have not attempted more than the briefest of summaries of what is a lively philosophical exchange. A conceptual pluralism may frustrate those who seek support for certain policies from a definition of human enhancement. Suppose that a Tour de France competitor who starts with normal levels of red blood cells takes synthetic EPO to shift his levels to a point slightly higher on the normal range. This is a case of enhancement as improvement. But, since the drug boosts his levels to a higher point within the normal range and not beyond it, the injections do not enhance beyond human norms. So is the competitor properly viewed as an abuser of performance-enhancing drugs? He has undergone enhancement as improvement but not enhancement beyond human norms. His levels of red blood cells are now high-normal rather than low-normal.

This dispute is a serious one for philosophers of enhancement who hope to address ethical problems arising on the problematic borderline between therapy and enhancement beyond human norms. But it will not concern us here. The cases that concern us involve improvements well beyond what is normal for human beings. Improvements of significant attributes and abilities to levels that greatly exceed what is currently possible for human beings are enhancements according to both definitions of human enhancement. One further justification for saying no more about how to define human enhancement is that a modification's status as an enhancement plays no direct role in this book's rejection of it. What is key is not that a modification is properly counted as a human enhancement, but rather its status as a *radical* human enhancement. The radicalness of some enhancements is what makes them transformative changes.

The Objective Ideal of Human Enhancement

According to the objective ideal, human enhancements have value commensurate with the degree to which they objectively enhance our capacities. We can represent the objective ideal graphically, as shown in figure 2.1.

The graph represents degrees of enhancement both positive and negative. The notion of negative enhancement seems an oxymoron—perhaps a better term would be *pejoration*, from the Latin *pejor* meaning "worse." The graph maps enhancements and pejorations of human capacities onto degrees of prudential value. Pejorations of a prudentially valuable capacity—such as intelligence or life span—tend to reduce that capacity's

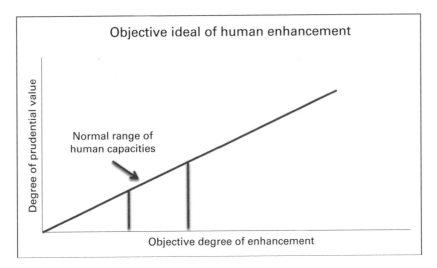

Figure 2.1

prudential value. They make the capacity less effective at promoting its possessor's interests or well-being. According to the objective ideal, enhancements of a prudentially valuable capacity tend to increase that capacity's
prudential value. They tend to make it more effective at promoting its possessor's interests or well-being. The objective ideal insists that the degree to
which our capacities make our lives go better increases with the objective
degree of their enhancement.

The graph depicts a particularly straightforward way of assigning degrees
of prudential value to objective degrees of enhancement. What is essential
to the objective ideal, as I understand it, is not the particular slope of the
line that links enhancement to value. The objective ideal corresponds to a
variety of functions that map degrees of enhancement onto degrees of prudential value. Some presentations of the objective ideal assign diminishing
marginal value to enhancement. It is often right to assign diminishing marginal utility to additional sums of money—your first million dollars is likely
to be more valuable to you than your second million. The first million opens
up a very large number of meaningful options—the ability to buy previously
unaffordable automobiles, to pay off debts, and to vacation at expensive
holiday resorts. The second million opens up comparatively fewer options.
The same pattern should obtain in respect of enhancement. Consider the
enhancement of human cognitive powers. Supposing that IQ is a genuine

measure of intelligence, the first 50 IQ points added by the application of a given technology could be more valuable than the additional 50 IQ points added by a second application. You are probably aware of many things that might be opened up by the addition of the first installment of IQ points—learning to play better chess or to read *War and Peace* in the original Russian, and so on. You probably have fewer occurrent desires corresponding to the activities enabled by the second 50 IQ point improvement. A graph that depicts this relationship between enhancement and prudential value would tend to become less steep with increasing degrees of enhancement. A shared feature of functions comprising the objective ideal is that they assign greater degrees of prudential value to greater degrees of enhancement. The objective ideal *never* assigns negative marginal value to enhancement—an additional increment of enhancement *never* reduces the prudential value of a capacity. Outcomes in which additional enhancement reduces prudential value are a feature of the anthropocentric ideal, which I describe later in this chapter. These claims about value implied by the objective ideal have implications for prudential rationality. All else equal, it is prudentially irrational to prefer a lesser degree of enhancement to a greater degree.

We can measure degrees of objective enhancement in a number of ways. Consider the following objective measure of the degree to which human cognitive powers could be enhanced. Computer scientists measure computing power in terms of the number of instructions a device can perform per second. Hans Moravec applies this measure to the human brain, arriving at an estimate of 100 million mips (where a mip is a million instructions per second) for the brain's 100 trillion synapses.[6] Suppose that we accept this as one objective measure of the current performance of the human brain. A modification that enabled the brain to perform 1,000 million mips would produce a ten-fold objective enhancement of human cognitive ability according to this measure. Consider another objective measure of human ability. The fastest human sprinters cover 100 meters in just under 10 seconds. Cybernetic implants that propelled them over the same distance just short of 5 seconds would result in a two-fold objective enhancement of their sprinting capacities. A collection of pills that increased human life expectancy from 80 years to 800 would produce a ten-fold objective enhancement.

The objective ideal requires that, all else equal, we prefer a genetic modification that boosts our brains' computing power to 1,000 million mips to a modification that boosts its power to 500 million mips. It directs that we prefer, all else equal, a cybernetic modification that enables us to cover 100 meters in 5 seconds to a modification that enables us to cover the same distance in 7 seconds. It condemns as prudentially irrational the selection of a life expectancy of 800 years when one of 1,000 years is available at no extra cost and with no additional side effects.

These are three of many objective measures of the enhancement of human cognitive abilities, physical prowess, or health. Objective enhancement is not just objective change; one does not undergo objective enhancement by eating sufficient pizza to increase one's weight from normal to morbidly obese. Objective enhancements should be viewed as the subset of objective changes that correspond to our prudential values. Nick Bostrom assembles a list of human attributes as candidates for objective enhancement. This list includes "healthspan," interpreted as "the capacity to remain fully healthy, active, and productive, both mentally and physically"; cognition, which comprises intellectual capacities such as "memory, deductive and analogical reasoning, and attention, as well as special faculties such as the capacity to understand and appreciate music, humor, eroticism, narration, spirituality, mathematics"; and emotion, "the capacity to enjoy life and to respond with appropriate affect to life situations and other people."[7]

There's no necessary synchrony or synergy between different varieties of objective enhancement. Consider human intellectual abilities: the digit span test is one way to measure the limits of working memory. Participants in the test hear sequences of digits that they are then asked to repeat. A person's digit span is the longest series of digits that he or she can accurately report. Perhaps there are aspects of a person's working memory that are not measured by his or her performance in a digit span test. It may nevertheless be an accurate measure of some aspects of working memory. A given modification may increase maximum digit memory four-fold while having no effect on other aspects of cognitive performance. It may even make performance in these other areas worse. According to some reports, the drugs Adderall and Ritalin, popular among students cramming for their final exams, may enhance powers of concentration at the same time as reducing creativity.[8] The objective ideal is compatible with a variety of views about the relative value of the enhancement of different capacities. In itself, it offers no guidance on how one might trade off the enhancement of digit span memory against a pejoration of some dimension of creative thinking, for example.

The objective ideal seems to support many of the things that transhumanists say about human enhancement. According to the influential document "The Transhumanist FAQ," transhumanism is an "intellectual and cultural movement that affirms the possibility and desirability of fundamentally improving the human condition through applied reason, especially by developing and making widely available technologies to eliminate aging and to greatly enhance human intellectual, physical, and psychological capacities."[9]

The most vivid account of how the objective ideal could offer practical guidance to the development of enhancement technologies comes from Ray Kurzweil. Kurzweil predicts that improvements of information processing technologies will lead humans to fuse with machines. Humans will progressively replace computationally cumbersome, disease-prone neurons with super-efficient electronic chips. According to Kurzweil, converting the human mind into technology will soon generate intellects "about one billion times more powerful than all human intelligence today."[10]

Kurzweil describes a series of technological revolutions in human enhancement. The genetic revolution will enable us to improve on nature's programming of human beings by modifying our DNA. The nanotechnological revolution will see the introduction of miniature robots into our bodies and brains. The robotic revolution will see the ongoing retirement of fallible, computationally cumbersome biology. Artificial computational devices will progressively replace biological brains—thinking devices whose design and programming was largely complete by the end of the Pleistocene era. According to Kurzweil, technological change will be so rapid that new technologies will succeed older ones almost instantaneously. We will soon arrive at *the Singularity*—"a future period during which the pace of technological change will be so rapid, its impact so deep, that human life will be irreversibly transformed."[11] Kurzweil offers 2045 as the year of the Singularity.

Kurzweil points to a number of examples of technologies that conform to this pattern of increasing value with technological progress. There is general agreement about what kinds of changes to laptop computers and mobile phones enhance them. Better computers have faster chips, bigger hard drives, and so on. They become correspondingly better at doing what their owners want them to do. While considerations of price often direct people to buy computers with less powerful chips than others that they might have bought, it is difficult to imagine making a case for the superiority of a chip to another that performs twice as many instructions per second but is identical in all other respects. Kurzweil's book *The Singularity Is Near* is replete with diagrams that show objective improvements in the power of our technologies. The objective ideal offers fully satisfactory descriptions of the value of improvements of dynamic random-access memory (RAM),[12] microprocessor clock speed,[13] the power of supercomputers,[14] the cost in sequencing DNA,[15] Internet data traffic,[16] and so on.

Kurzweil urges that we take the same approach to human mental and physical capacities and technologies that could enhance them. He gives many examples of technologies whose value increases with enhancement. The enhancement of our mental and physical capacities should, according to Kurzweil, follow the same pattern. The same law of accelerating returns that leads to better mobile phones will also create better human minds.

Must transhumanists say that objectively greater degrees of enhancement of prudentially valuable capacities are *always* more valuable than objectively lesser degrees? Are they committed to the objective ideal? The notion that they are is strongly supported by an explicit transhumanist rejection of the relevance of facts about current human capacities or interests to the prudential value of degrees of enhancement. Transhumanists reject the relevance of appeals to human nature or to facts about what humans typically do. The transhumanist thinker Max More conveys the irrelevance of such appeals in a particularly vivid way. He writes a "Letter to Mother Nature."[17] The letter expresses a somewhat theatrical gratitude to Mother Nature for having "raised us from simple self-replicating chemicals to trillion-celled mammals" and having given us "a complex brain giving us the capacity for language, reason, foresight, curiosity, and creativity." But it's not all good. More thinks it's time for humans to take up where Mother Nature has left off. We must amend the human constitution to radically extend our life spans and enhance our cognitive powers. "These amendments to our constitution will move us from a human to an ultrahuman condition as individuals. We believe that individual ultrahumanizing will also allow us to form relationships, cultures, and polities of unprecedented innovation, richness, freedom, and responsibility." More's letter conveys an attitude toward Mother Nature and human nature reminiscent of the attitude of the dolphins in Douglas Adams's *Hitchhiker's Guide to the Galaxy* series who, upon learning that the Earth is to be destroyed depart the planet, leaving humans a note "So long, and thanks for all the fish." If we have to become something other than human—perhaps posthuman or ultrahuman—to enjoy the increasing benefits of enhancement, then so be it.

The Instrumental *and* Intrinsic Value of Human Capacities

I propose a disanology between the enhancement of human capacities and the enhancement of technologies that humans use.

We view technologies as *instrumentally* valuable—valuable because of what they enable us to do. It's difficult to think of another significant contributor to the value of laptop computers or mobile phones. Some mobile phones are more aesthetically pleasing than others, but this does not seem to be a significant contributor to judgments about which of two phones is better. The prettiness of an Apple iPhone would not long preserve its market share were its competitors to sport better access to wireless networks, sharper displays, greater data storage capacity, cameras with more megapixels, and so on. An Intel engineer who tries to make a case for a new computer chip that is identical to an existing one in all functional respects apart from the fact that it performs fewer calculations per second faces a disappointing Christmas bonus. His bosses will rightly doubt that there could be any significant alternative contributor to the value of a computer chip. This instrumental value conforms to the objective ideal.

Human capacities differ. The difference does not lie in the irrelevance of the instrumental approach to enhancements of our capacities. They are instrumentally valuable. Stronger muscles are more instrumentally valuable in part because they enable us to move heavier objects. The enhancement of our cognitive capacities increases their instrumental value by enabling us to solve more difficult problems. We convey the plight of people with physical or cognitive disabilities by describing the reduced instrumental value of their capacities. A severe spinal injury reduces the instrumental value of legs that no longer enable their possessor to walk. A cognitive disability reduces the propensity of cognitive capacities to solve problems. Kurzweil's law of accelerating returns could describe great improvements of our capacities' instrumental value. The difference between human capacities and technologies used to extend human capacities lies in the existence of a significant alternative contributor to the value of the former that does not apply to the latter. This alternative contributor to prudential value does not conform to the objective ideal. According to what I will call the anthropocentric ideal, some objective enhancements reduce the prudential value of human capacities.

We get an incomplete picture of the value of enhancing our capacities if we rely exclusively on the objective ideal to assign value. The objective ideal does not describe all of the effects of the enhancement of human capacities on prudential value. There is another way to value human capacities that recommends a different way to value their enhancement.

Anthropocentric Ways of Evaluating Enhancements

The anthropocentric ideal assigns value to enhancements relative to human standards. Some enhancements of greater objective magnitude are less valuable than enhancements of lesser magnitude. We can represent this ideal graphically, as in figure 2.2.

I do not intend to reject the objective ideal in favor of some version of the anthropocentric ideal. Rather, I argue that the objective ideal describes changes to the instrumental value of human capacities. The anthropocentric ideal differs in describing different degrees of *intrinsic* value. We arrive at a final reckoning of the prudential value of an enhancement by

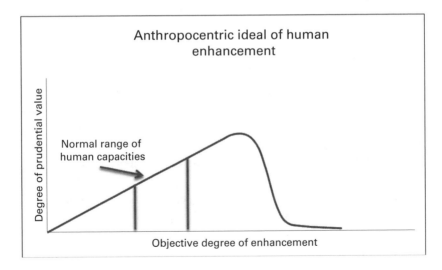

Figure 2.2

appropriately balancing the instrumental and intrinsic value of the resulting capacity.

In what follows, I propose that the instrumental value of a capacity corresponds to its propensity to yield *external goods*. A capacity's intrinsic value corresponds to its instantiation of *internal goods*. The term "intrinsic value" has a variety of philosophical uses.[18] I intend a quite specific meaning here. A capacity has intrinsic value through its instantiation of intrinsic goods. The term "instantiate" indicates a relationship between capacities and goods that differs from that obtaining between a capacity and separable external goods. I look to the philosopher Alasdair MacIntyre to clarify the distinction between internal goods *instantiated by* the exercise of capacities and external goods *resulting from* exercises of our capacities.

In a widely cited passage, MacIntyre uses the example of chess-playing to separate external from internal goods of exercises of our capacities. Chess's external goods connect to the playing of the game "by the accidents of social circumstance."[19] One may acquire the external good of prize money by beating opponents in a chess tournament. But this connection is contingent. It is not absolutely essential to play the game to get the prize money—a more effective way might be to break into the venue hosting the tournament and thieve it. Many chess tournaments lack cash prizes. It's possible to play in exactly the same way against the same opponents but receive no financial reward. The internal goods of chess are not separable from the activity of playing chess in the way that external goods are. MacIntyre says that "they cannot be had in any way but by playing chess or some other game of that specific kind." Furthermore, the internal goods of chess "can only be identified and recognized by the experience of participating in the practice in question,"[20] in this case by actually playing chess—or some other game of that specific kind. One acquires chess's internal goods in pondering alternative endgame strategies, evaluating a possible sacrifice of your queen, and staring down an opponent across the chessboard. These are not separable from the activity of playing chess in the way that the game's external goods are.

MacIntyre argues that internal goods have been systematically underrecognized and underappreciated. Contemporary liberal democratic political arrangements are, according to him, overly focused on external goods. I am not interested in MacIntyre's attempt to revive virtue ethics here. I assume that human capacities have value by virtue of connections with both external and internal goods. I assume no view about the relative significance of internal and external goods, and no view about their overall contribution toward human individuals' overall levels of well-being.

A capacity has instrumental value in terms of its propensity to yield external goods and intrinsic value in terms of its propensity to instantiate internal goods. Both kinds of value come in degrees. We can assign differing degrees of instrumental value in accordance with a capacity's facility in providing external goods. A capacity is more instrumentally valuable if it furnishes better or greater numbers of external goods. We can assign differing degrees of intrinsic value in accordance with a capacity's facility in instantiating internal goods. A capacity is more intrinsically valuable through instantiating more valuable internal goods.

The two types of value respond differently to enhancement. As we have seen, a capacity's instrumental value tends to increase with the degree of its objective enhancement. Our capacities share this tendency with technologies. Returning to MacIntyre's chess example, the enhancement of one's ability at chess should enable one to win tournaments with larger cash prizes. The better one is at playing, the greater the social recognition one tends to receive. Some external goods from playing chess have an upper limit. There's a large but finite amount of prize money available for chess tournaments. But other external goods from enhancements do not seem to have obvious upper limits. There should always be heavier loads that additional enhancements of strength will enable you to lift, and always more difficult problems that additional cognitive enhancements will enable you to solve.

In chapter 3, I argue that the internal goods of our activities respond differently to enhancement. They tend to increase up to a certain point. Beyond a certain point, however, the quality and quantity of these goods tend to decline. If we assign intrinsic value to a capacity according to the internal goods it yields, then this value will be independent of any external goods that exercise of a capacity tends to grant us. An enhancement may greatly increase a capacity's instrumental value while significantly reducing its intrinsic value. The intrinsic value of human capacities conforms to the anthropocentric ideal. A great degree of objective enhancement of chess-playing ability increases pleasures inherent in playing the game. Beyond a certain point, however, the enhancement of our chess-playing ability does a worse job of instantiating chess's distinctive internal goods. It therefore makes the mental abilities that chess engages less intrinsically valuable.

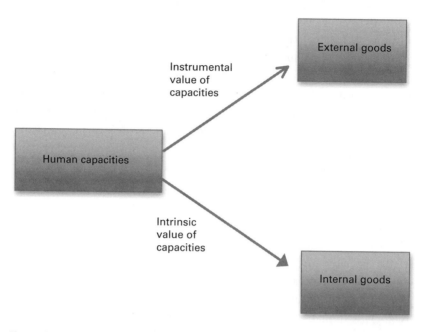

Figure 2.3

This is not to deny that radically enhanced minds won't seek out their own internal goods through the playing of chess—possibly a multidimensional version of the game. The anthropocentric way of valuing enhancement allows that humans who undergo radical enhancement may come to adopt a different view of enhancements once they have received them. The idea that radical enhancement alters our evaluative framework suggests the possibility of divergent views about the transformed state. A result that you rightly fear now could be the occasion for justified celebration once achieved. Remember that these divergent verdicts are a feature of the transformative change of body-snatching. Human individuals rightly fear being turned into pod-people. However, once transformed, they are rightly happy to have undergone it. They will recognize that, but for the transformation, they would be denied experiences and achievements that they rightly value. A species-relativism about prudential value denies the possibility of a species-independent standard from which to assess human existences as either superior or inferior to pod-person existence. Both humans and pod-people are rational beings who rightly value aspects of their lives not shared by members of the other group.

As it is with body-snatching, cyberconversion, and Borg assimilation, so it is with radical enhancement. Radical enhancement may turn us into beings who are grateful for having undergone the transformative change. It is, nevertheless, in conflict with the evaluative framework human beings apply to enhancements. It disconnects us from distinctively human internal goods.

Chapters 3 and 4 explore the phenomenon of alienation by radical enhancement. They sample two varieties of experience and achievement greatly valued by human beings. While this book cannot pretend to cover all humanly valuable experiences and achievements, I propose that what goes for these two experiences and achievements also holds for many others.

Chapter 3 explores, in some detail, the value that superhuman feats have for us. We can imagine enhancements of our sprinting prowess that would make Usain Bolt look pedestrian, enhancement of certain of our cognitive powers that would make Albert Einstein seem an unimaginative plodder. My anthropocentric response to these possibilities focuses on the experiences they enable. I argue that we value possible experiences within and slightly beyond the normal human experiential range because we humans can engage with them veridically—when we imagine what it is like to have these experiences we can represent them to ourselves approximately as they are. This variety of veridical engagement is what we are unable, or less able, to do when we seek to imagine ourselves as a panda eating some bamboo or as a computer successfully executing a virus-checking program. We may not always succeed in our attempts to engage veridically with other humans, but success seems possible. It's the possibility of veridical engagement that makes possible future experiences meaningful to us as we are now. Possible experiences far outside of the normal human range are less meaningful for us because we're less able to engage veridically with them. We might become the kinds of beings who would view Bolt's performances as pedestrian and Einstein's insights as obvious. But the fact that we could become such beings does not require us to positively anticipate the changes to our values.

Chapter 4 switches the orientation of these concerns about the value of experiences. What are the consequences of radically enhanced experiences and achievements for those who acquire them? I explore a danger for the identities of individuals who undergo radical enhancement. Radical enhancement may have the effect of bringing human lives to premature ends.

In chapter 5, I explore the effects of cognitive enhancement on our acquisition of knowledge. Cognitive enhancement promises to dramatically extend our understanding of the universe and our place in it. I argue that the advance of science is subject to an anthropocentric constraint. We place a reduced value on the new knowledge that radical cognitive enhancement would bring.

Beyond a certain point, enhancement tends to reduce the intrinsic value of human capacities. As it has this effect, it boosts their instrumental value. How are we to choose between two alternatives—a level of capacities that combines high intrinsic value with low instrumental value and a level that combines low intrinsic value with high instrumental value? In chapters 3 and 5 I argue that we may not need to choose. The enhancement of human capacities is not an effective way to pursue instrumental value. If we ask the question prospectively—which interventions are likely to boost instrumental value?—we find that we can preserve our capacities' intrinsic value while doing our best to enhance our access to external goods. I will argue that we should limit instrumental enhancement to technologies that we use to extend human capacities.

3 What Interest Do We Have in Superhuman Feats?

The first years of the twenty-first century have seen a resurgence of the Hollywood superhero movie. A spate of big-grossing movies celebrates humans or humanlike beings with superhuman abilities. For example, the 2012 movie *The Avengers* features flying humans, humans with super-strength, humans sufficiently robust to survive skyscraper collapses, humans with the physical agility to dodge bullets, and humans with exceptional mental powers. Does the popularity of these movies suggest a perhaps only partially acknowledged desire to radically enhance our physical or mental powers? Should we pursue superhuman feats by means of genetic or cybernetic technologies?

An interest in superhuman feats seems to conform to the objective ideal. The greater the degree of enhancement enabled by the application of genetic or cybernetic technologies to our bodies and minds, the more impressive will be our physical and mental feats. One way in which we value our capacities does conform to the objective ideal. Radical enhancement greatly increases the instrumental value of human capacities. There is, however, another kind of value that we attach to our capacities which follows the anthropocentric ideal. This mode of evaluation assigns increasing intrinsic value to enhancements of our capacities across and somewhat beyond the normal human range. Somewhere beyond that range it assigns decreasing intrinsic value.

Radical enhancement brings the objective and anthropocentric ideals into conflict. Gains in instrumental value seem to entail sacrifices of intrinsic value. Perhaps we do not, as a practical matter, have to resolve this conflict by selecting either the objective or the anthropocentric ideals to guide our deliberations about enhancement. There are ways to procure many of the goods that track the objective ideal that do not involve enhancing human capacities. They are likely to be more effective at providing many of the goods that motivate the radical enhancement of human capacities. Furthermore, they do not reduce our capacities' intrinsic value.

The Value of Enhanced Marathons

Haile Gebrselassie won the 2008 Berlin marathon in a world record time of 2 hours, 3 minutes, 59 seconds. His achievements have prompted intense debate in distance running circles about whether or when a human marathoner will run a sub-two-hour marathon. Here's something that participants in this debate *don't* consider themselves to be talking about. They don't consider themselves to be debating whether the application of a genetic or cybernetic technology to an athlete might result in sub-two-hour marathons. The answer to this question is a pretty straightforward "yes."

Considered in objective terms, Gebrselassie's performance is actually pretty mediocre. You would be unlikely to purchase a used car whose best performance over 42.195 kilometers was 2 hours, 3 minutes, 59 seconds. There are likely to be many ways of combining human brains and bodies with technology that would lead to marathons objectively superior to Gebrselassie's. Consider a possible future technology that would radically enhance the endurance of human athletes and would enable marathons to be completed in times dramatically faster than 2 hours. One of the main limitations on the performances of competitors in the marathon is the amount of oxygen that can be carried to competitors' muscles. Human athletes rely on hemoglobin to do this job. Injections of a synthetic version of the hormone erythropoietin (EPO) boost a competitor's supply of hemoglobin. Consider one way in which the endurance of human athletes could be not only enhanced, but radically enhanced. Robert Freitas has described something he calls a respirocyte, a one-micron-wide miniature robot, or nanobot, designed as a replacement for hemoglobin.[1] Respirocytes could boost endurance to levels far beyond those enabled by injections of EPO. They are currently mere theoretical entities. But should Freitas's dreams be realized, respirocytes could be introduced into human bodies, dramatically outperforming hemoglobin in keeping our tissues oxygenated. Respirocytes would carry approximately 236 times the quantity of oxygen carried by the hemoglobin they replace. According to Freitas, a few cubic centimeters of them could "exactly replace the gas carrying capacity of the patient's entire 5.4 liters of blood." He imagines that respirocytes could "enable a healthy person to sprint at top speed for at least 15 minutes without breathing, or to sit underwater at the bottom of a swimming pool for hours."[2] Such enhancements of endurance fit squarely within the category of radical enhancement. If Kurzweil and Freitas are right, these enhancements enable feats that are significantly more impressive than those attainable by unenhanced humans.

We appear to combine an intense interest in the athletic achievements of unmodified human competitors with a relative indifference to the possible achievements of enhanced athletes. In his book *Enough*, Bill McKibben seeks to explain and justify a lack of interest in the faster marathons potentially permitted by the modification of competitors' genomes. He thinks that such enhancements would make marathons plain boring. According to McKibben, if marathoners genetically enhance themselves, "we won't just lose races, we'll lose racing: we'll lose the possibility of the test, the challenge, the celebration that athletics represents." This is because genetic enhancement transforms the marathon from an event in which competitors are pushed to their physical and mental limits into one in which they perform according to their design specifications. According to McKibben, running would become "like driving." He allows that "driving can be fun," but insists that "the skill, the engagement, the meaning reside mostly in those who design the machines."[3] This last claim might be news to Michael Schumacher.

The suggestion that there will be no challenges for genetically modified athletes confuses enhancement with omnipotence. There will be challenges for genetically enhanced athletes. If we allow genetic enhancement, we won't "lose racing." We *may* lose some of the races that humans currently compete in. If 42.195 kilometers doesn't offer genetically enhanced long distance runners a proper challenge, they'll only run it as part of a much longer distance—perhaps as part of a posthuman marathon of 421.95 kilometers.

McKibben imagines genetically enhanced marathoners winning without having to try hard. Winning will require nothing more than the correct operation of competitors' internal accelerators. This is unlikely. In biological terms, trying hard is essentially the allocation of additional resources toward an activity deemed especially important. This is what human marathoners are doing when they put everything they have into sprinting over the final 100 meters. Cars can't try hard. They have no capacity to decide to transfer resources normally required for functions such as maintenance of the air conditioning system or the operation of the brake lights, to boosting speed. Posthuman marathoners who are, like cars, incapable of trying hard will probably lose to those who are capable of mobilizing additional resources. The winners of posthuman marathons may even try harder than the winners of human marathons. That is, they may have an even greater capacity to channel resources normally required for other bodily or mental functions into long-distance running performances.

Of course, these superhuman levels of effort may only be possible because of prenatal genetic modifications. But posthuman marathoners can acknowledge this fact at the same time as taking pleasure in and accepting credit for their achievements. Their attitude might resemble that of a God-fearing gold medalist who can acknowledge a debt to the creator at the same time as taking pride in victory. One thanks God, while the other thanks her genetic engineer. In both cases competitors take pride in turning what they were born with into exceptional performances. No marathons, human or posthuman, will be won by simply turning up at the starting line with certification of genetic enhancement.

Simulation and Meaning

I propose that our preference for Gebrselassie's marathon over the objectively superior performances of enhanced runners accords with criteria that are parochial but principled. It reflects something deep about human psychology. We find Gebrselassie's marathon exploits more valuable than we would the more objectively impressive achievements of nanotechnologically enhanced athletes and second-hand cars because of an access we have to the former that we lack to the latter. Human spectators can engage with the former in a way that they cannot with the latter. The possibility of engagement with the performances of others and our own future performances supports the anthropocentric pattern of value. It corresponds with a capacity to engage veridically with these performances.

By veridical engagement, I mean that we can accurately imagine ourselves doing as the competitors do. There's a distinctive thrill from actually seeing Usain Bolt run 100 meters in 9.58 seconds that we do not experience through merely being told that he covered the distance in this time.

Exposure to the full set of movements of Bolt's body invites us to imagine doing as he does. Footage of dung beetles heroically wrestling with unfeasibly large (for them) balls of excrement may prompt similar efforts at imaginative identification. But unlike our engagement with Bolt, this engagement is not veridical. Our imaginings will not correspond to anything occurring in the head of the dung beetle. We should not claim any genuine insight into the beetle's experiences.

Veridical engagement comes in degrees. The extreme athletic prowess of Usain Bolt poses difficulties to human observers who hope for veridical insights into his psychology. My listless Saturday morning jogs up and down the hills of Wellington are likely to be more straightforward imaginative targets for human observers. Bolt's performances attract bigger audiences than my more easily imagined efforts because his are simply more exciting to engage with. Bolt's humanity encourages us in the belief that we can engage veridically with his exertions.

I think we can make sense of some complaints about performance-enhancing drugs. They make athletes' performances more objectively impressive. But at the same time, they open up a gap between athletes' performances and spectators, most of whom do not take anabolic steroids.[4]

These claims stand in need of an account of veridical engagement. I appeal to a psychological theory—simulation theory—to play this role. It enables us to separate experiences with which we can engage veridically from experiences that are beyond our powers of veridical engagement. Here's an explanation for our ready engagement with human marathoners and our comparative disengagement with machines and enhanced humans that seem, according to the objective pattern of value, to run better marathons.

According to simulation theory, we predict and explain the actions of other human beings by simulating the mental process behind them.[5] The results of this simulation enable us to predict what they will do and to explain why they did what they did. We predict what people will do when confronted by misdirected buses by imaginatively placing ourselves in those circumstances. In effect, we use our own psychological processes to simulate theirs. We predict that they would leap out of the way because that's what we would do in those circumstances. When we perform our simulations, we take our mental machinery "off-line" so that it produces predictions and explanations of others rather than causing us to act in certain ways. In simulating the mental processes of the person in the path of the bus we understand that *she* must take evasive action; the output of the simulation isn't a desire that *we* abruptly throw ourselves toward the curb.

Recently, simulation theory has found support in discoveries of mirror neurons, neurons that fire both when certain actions are performed and when these actions are observed or anticipated in others.[6] Simulation theory is not immune to refutation. But the reality of our emotive and psychological anthropocentrism survives the falsehood of this particular explanation for it. My argument goes through if our evaluative anthropocentrism is no ephemeral consequence of an ideology opposed to human improvement but rather is an enduring and robust feature of the human capacity to value experiences and achievements.

Simulation theory may explain our success in predicting the behavior of other human beings, but does it have implications for our assignments of value? A car can cover 42.195 kilometers faster than Gebrselassie. Its performance is therefore objectively superior. Why should the fact that we more readily simulate Gebrselassie's performance than the car's mean that we can place greater value on Gebrselassie's exploits than on those of a car?

Gregory Currie's account of our experience of reading fiction is an example of how our simulations might guide value assignments.[7] Currie claims that we simulate when we read a work of fiction. We engage emotionally with the pleasures and pains of fictional characters by simulating their mental states. According to Currie, "if our imagining goes well, it will tell us something about how we would respond to the situation, and what it would be like to experience it."[8] We enjoy reading Charles Dickens's novel *Great Expectations* because of our access to the aspirations, fears, successes, and disappointments of Pip, an orphan making his way in Victorian England.

Simulation theory assumes sufficient similarity between the individual doing the simulating and the individual being simulated. The explanations and predictions that emerge from our simulations depend on shared capacities and limitations. Currie's theory explains a bias toward human characters in works of fiction enjoyed by human readers, watchers, or listeners. It's notable that any nonhuman characters in fiction—Richard Adams's rabbits in *Watership Down*, the toys in the *Toy Story* movie franchise, and so on—have unrealistically humanlike psychologies. Human readerships and audiences would not find works of fiction populated with realistic rabbits and toys particularly interesting. This is not to deny that realistic rabbits and toys can be interesting; rather it is to say that they are not well suited to central roles in fictions read, listened to, or watched by human beings.

I propose that simulation theory offers a convenient description of our engagement with experiences that enhancement might enable. It explains the pleasure we get from achievements toward the top of the human range. We enjoy watching Gebrselassie's marathon performances because we can vicariously experience them. We can extrapolate from our own experiences of completing ten-kilometer jogs or of successfully running to catch buses. We thereby sample Gebrselassie's joy and exhaustion as he crosses the finish line. But there are degrees of athletic enhancement that push beyond our powers of veridical engagement. Outside of the human range, there's a point at which objective improvements become less interesting and therefore less valuable to us. This is because too great a degree of enhancement weakens our connection with the exploits it enables. The DC Comics hero Flash, a fictional crime-fighter possessed of the attribute of super-speed, is apparently capable of running faster than the speed of light. To the extent that fans do engage with him they likely imagine his athletic exploits in only vague and approximate ways. We humans have no opportunity to engage veridically with light-speed marathons, even if they are physically possible.

So far, my examples have involved different ways to complete marathons. Analogous points apply to MacIntyre's example of chess. One of the more significant events in recent chess history was the 1997 defeat of Garry Kasparov, possibly the strongest ever human chess player, by the IBM-designed computer Deep Blue. Over the course of the match, Deep Blue's movement of its chess pieces was objectively superior. Can we defend a preference for Kasparov's objectively inferior play?

Leon Kass and Eric Cohen address the problem of chess computers that appear to play chess even better than the best human players. They insist that chess will continue to be a game in which one human plays another in an era of apparent computer supremacy over the chessboard because "only we can play chess, that is, as human beings, as genuine chess players." Computers do nothing more than "play" chess. According to them, that is why no one would "watch a 'match' between two chess-playing computers." It's also the reason no one would watch "a baseball game that pitted robot pitchers against automatic batting machines."[9]

But we should be suspicious of arguments that restrict chess (or baseball) to human beings. Robot pitchers and automatic batting machines aren't playing baseball simply because it takes more to play baseball than pitching or batting. A genuine player of baseball who has just pitched a ball stands ready to take a return catch. A genuine player who has just hit the ball is launching him- or herself toward first base. The fact that baseball is more than pitching and batting renders machines limited to either pitching or batting therefore incapable of playing the game much in the way that someone whose knowledge of chess is limited to the correct movement of the knight is incapable of playing the game. Kass and Cohen are guilty of an illegitimate speciesism. Bonobos taught the rules of chess should lose the majority of the games they play because they're not very good players, not owing to disqualification on the grounds of species-membership. Computers moving chess pieces around the board in ways that deliberately conform to the rules of the game with the aim of checkmating an opponent are playing chess, not merely "playing" it. The key difference between Deep Blue and the bonobo player is simply that the computer is better. It's good enough to beat the best humans.

Kass and Cohen credit the designers of chess programs with the desire to build something able to "'play' perfect chess." This, they suggest, is not what human chess players are trying to do. Kasparov would, doubtless, love to play perfect chess. But his goal in preparing for his 1985 match against then-world champion Anatoly Karpov was simply to win. The IBM programmers of the chess computer Deep Blue had a similar goal when preparing their charge for the 1997 match against Kasparov. They wanted to win. Perfect chess, whatever that is, would have been a wonderful achievement, but all they really needed was for Deep Blue to play better than Kasparov.

I think Kass and Cohen are right to say that we will maintain our interest in human players even after chess programs routinely beat the best humans. This makes sense. We enjoy watching human 1,500-meter runners even though we know that any used car lot contains machines that beat them with ease. Kass and Cohen observe that "the computer 'plays' the game rather differently—with no uncertainty, no nervousness, no sweaty palms, no active mind, and, most crucially, with no desires or hopes regarding future success."[10] Their point survives removing the scare quotes around "plays." Our simulations include the fear that our opponent might have spotted a flaw in our cunning plan, and joy when it/he/she makes the moves that we predicted. We like watching the games of individuals who genuinely feel these things. Chess players who do not have these experiences are not particularly interesting objects of imaginative identification.[11]

No matter how soundly Deep Blue beats Kasparov, a human player will always play chess in ways that interest human spectators to a greater degree than Deep Blue and its successors. Human chess players of modest aptitude can read Kasparov's annotations and thereby gain insight into his stratagems. Kasparov's chess play is vastly superior to that of his fans. But he, presumably as a very young player, passed through a stage in his development as a player when his play was, in many respects, similar to that of the fans. There seems, by contrast, to be a barrier between human minds and vicarious experience of Deep Blue's play. According to its programmers, Deep Blue's chess brain contained a database with around 700,000 games. Each of these could be retrieved in their entirety within an instant. Deep Blue could generate positions and evaluate them using a function integrating four distinct measures of chess success at a speed of 200,000,000 positions per second.[12]

It's not that we couldn't attempt to play this way. Deep Blue's programmers almost certainly tested its algorithms in their heads, playing through short sequences of moves, as a first check of their viability. A human who attempted to implement Deep Blue's algorithms, reducing the computer's rate of generating and evaluating positions from 200 million per second to around one position every five minutes, and the size of its database from 700,000 to a small number of partially remembered games, might find himself having to explain how he managed to lose to a chess-playing bonobo.

Perhaps Deep Blue plays better chess than Kasparov. But Kasparov plays the game in a way that we can vicariously experience, enjoy, and therefore value. We can engage veridically with the play of Kasparov, but not with that of Deep Blue.

Analogous points apply to chess victories that we might achieve through chess-specific cognitive enhancement. Those who are most impatient to radically enhance their cognitive abilities look with optimism toward advances in artificial intelligence. Advances in computing power will, they think, soon yield machines with the computing power of the human brain. These advances will give rise to technologies that enable functions of the human brain to be taken over by electronic neuroprostheses that perform these tasks to a much higher standard. Why bother using computationally clumsy neurons to do long division when you might hand the task over to a brain implant built for the purpose of mathematical calculation? This may also be the path to enhancement of our abilities as chess players. Suppose that all of Deep Blue's chess knowledge and skill could be loaded onto a chip that was then appropriately implanted in your brain. Your chess skills would have undergone enhancement. The procedure has increased the instrumental value of your chess play. Supposing that chess officials do not ban you from competition, you achieve many external goods. You win tournaments and garner cash prizes. You should see the enhancement as decreasing the intrinsic value of your chess playing. It has reduced your access to the internal goods that partially motivate you to play the game. The transformative change effected by inserting the chip is likely to place new internal goods within your reach. Those who choose to steal a chess tournament's prize money rather than participating in the event connect with the internal goods of breaking and entering. But these are not the internal goods toward which human chess play is oriented.

Our simulations are the basis of our connection with the performances of others. They explain our interest in and valuing of those performances. But they're also the basis of our valuing of our own future performances. Here the relevant evaluative connection is not with others, but instead with our own future selves. When we think about things that we might do, we place greater value on achievements that are within or just beyond the normal human range. It's only within and slightly beyond this human range that we place systematically greater value on more objectively impressive achievements. Our anthropocentric norms acknowledge 2 hours 30 minutes as a better marathon time than 3 hours. We view solving a hard Sudoku puzzle as better than almost solving one of medium difficulty. The limits on our capacity to engage with various possible experiences have implications for the value we place on different degrees of enhancement. Within and perhaps somewhat beyond the normal human range, we place greater value on experiences that correspond with more objectively significant accomplishments. We can, of course, attempt to imagine what it would be like to complete the 100-meter dash or to solve hard Sudoku puzzles in 0.9 milliseconds. We might succeed. However, these objectively more impressive achievements are less meaningful for us. We're right to place a reduced value on possible experiences corresponding with accomplishments such as these that lie far outside of the normal human range.

We could undergo transformative changes that would lead us to value radically enhanced marathon or chess performances. I have appealed to simulation theory to explain a reduction in our ability to engage with radically enhanced performances. We struggle to simulate the chess play of Deep Blue and the foot races of the Flash because we fall far short of having cognitive and physical capacities to produce these performances. Radical enhancement would, presumably, give us the requisite capacities. It would grant us the distinctive collection of internal goods available to radically enhanced beings. But there is a cost. In doing so it deprives us of internal goods instantiated by our current chess playing and marathon running. When mediocre human chess players simulate the play of Kasparov, they experience it as valuable. We recognize Kasparov as doing a superior version of what we do. A player whose moves are guided by a chip running an upgraded version of Deep Blue's program has no such capacity. He views Kasparov's play as poor because of the evident quality gap between Kasparov's play and his own. Kasparov's ability to compute the value of all positions that may result from a given position does not come close to that of the chess chip. He may occasionally fail to respond to a position tagged in Deep Blue's memory as leading to a forced checkmate. The human player enhanced with the Deep Blue chip has a reduced ability to view as valuable his own play prior to the insertion of the chip.

Radical enhancement would not completely deprive us of an ability to appreciate our former achievements. But that appreciation is necessarily indirect. We can be impressed by the efforts of a dung beetle in an indirect way. "Such a small beetle, such a (relatively) large ball of dung!" The radical enhancement of our physical powers leaves us with the indirect mode of access to the achievements of our past selves. As it happens, most of us are familiar with this relationship with past experiences. As adults, we have this manner of access to many of our achievements as children. In chapter 4 I will discuss and reject an argument that infers the goodness of undergoing radical enhancement from the goodness of growing up.

The recipient of a Deep Blue chess chip has undergone a transformative change altering the value assigned to these different ways to play chess. We should not view these changes from the evaluative standpoint that we might adopt postenhancement. That would be like accepting the offer of the pod-person Dr. Kauffman to assess the prudential rationality of choosing to be snatched from the perspective of the pod-person you could become. There's no surprise that the pod-you celebrates the loss of love, desire, ambition, and faith. But that doesn't mean that you should. We should consider both chess enhancement and body-snatching from the evaluative perspective we possess when contemplating the transformation.

Is Human Enhancement the Right Way to Pursue External Goods?

I have argued that some of the value that humans attach to possible future experiences does not correspond with the degree of enhancement of the capacities that permit them. Beyond a certain point, enhancement enables experiences from which we are imaginatively disconnected. We assign lesser degrees of intrinsic value to them. This analysis omits reference to instrumental benefits that do seem to increase with the degree of enhancement. The instrumental benefits of enhancement conform to the objective pattern. Stronger people can carry heavier burdens. Even stronger people carry even heavier burdens. Those who run faster arrive at their destinations sooner. MacIntyre's chess example involves external goods that are not particularly socially valuable. Enhancements of chess play that bring greater cash prizes benefit the individuals who receive them without bringing broader social benefits. This risks giving a false impression of external goods. Other external goods are very socially valuable. Suppose that radical cognitive enhancement were to enable the discovery of a cure for cancer. If the argument given so far is correct, then the process of finding the cure would lack many of the valuable internal goods that tend to accompany our scientific discoveries. But we'd still have the cure. The cure would prevent a vast quantity of human suffering. Analogous points apply to other problems—climate change, global terrorism, wealth inequality, and so on—decisive solutions to which may be beyond our current intellectual powers. In such cases, the external goods of radical enhancement could be so valuable that it seems absurd to worry about the low intrinsic value of the activities that generate them.

We seem to face a dilemma. Either we insist that the influences directed at our brains and bodies preserve our association with distinctive human internal goods and thereby forgo the external goods that would come from the pursuit of enhancements with high instrumental value, or we renounce many distinctive internal goods as we attempt to boost the instrumental value of our capacities.

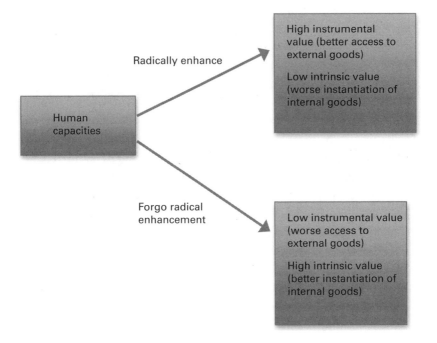

Figure 3.1

One option might be to argue that internal goods are systematically more important than external goods. As a general thesis, this seems somewhat suspect. We would surely choose to have the cure for cancer even if the process leading to it brought none of the internal goods associated with scientific discovery.

Perhaps the preceding paragraphs do not accurately represent the choice we have to make. Some defenders of enhancement like to call their philosophical and political opponents *bio-luddites*—a name adapted from that of the Luddites, a movement of early-nineteenth-century textile workers violently opposed to the Industrial Revolution. So-called bio-luddites oppose the genetic, cybernetic, and other technologies that would improve human beings. This charge of Luddism is misdirected. Opponents of radical enhancement need not be enemies of technology. Rather, they oppose a misuse of technology. Here I mean "misuse" not to indicate any moral judgment. In choosing to pursue external goods by enhancing human capacities, we forgo a more direct and efficient means of pursuing them.

The fact that the radical enhancement of our capacities increases their instrumental value does not show that a desire to enhance instrumental value should lead us to enhance our capacities. There are likely to be better ways to achieve the goods we seek. Here are two questions:

(1) Should we accept external goods were they to be delivered by radical enhancement?

(2) Should we pursue external goods by means of radical enhancement?

The answer to question (1) is sometimes yes. It would be absurd to reject a cure for cancer if it were achieved through the radical enhancement of our intellectual powers. We wouldn't reject a cure offered by extraterrestrial visitors. This would be so even if we regretted the lack of internal goods that accompany human scientific discoveries. Those who read Michael Bliss's book *The Discovery of Insulin* learn about a hugely important discovery. Readers also get a moving account of two human individuals, Frederick Banting and Charles Best, who struggled, frequently failed, but finally triumphed in their quest to isolate the hormone that was either absent from or present in insufficient quantities in the bodies of diabetics. Knowing about insulin is of such great value to us that we would clearly accept this knowledge even if it came without a moving story of human endeavor. By analogous reasoning, it would be absurd to reject a cure for cancer because it came by way of the enhancement of scientists' cognitive powers.

We can answer question (1) in the affirmative while insisting that the answer to question (2) is almost always no. The radical enhancement of our capacities is an inefficient way to pursue instrumental goods. Compare: You might be disposed to accept a confirmed cure for cancer discovered by astrological investigation even as you insist that astrology is a bad place to invest cancer research dollars. The enhancement of our human capacities—the option of internalizing enhancement—is a more effective means of getting science's external goods than is astrology. But it is a less efficient way of pursuing most of these goods than is the option of seeking to invent new nonhuman technologies to aid our investigations—the option of externalizing enhancement. The fastest route to a cure for cancer is unlikely to be by way of the enhancement of the cognitive capacities of medical researchers.

Consider a choice between internalizing or externalizing enhancement. Some of the larger Egyptian pyramids required tens of thousands of workers to commit themselves to the task of sculpting, hauling, and placing stone for a decade or more. The work was back-breaking—sometimes literally. There is a clear need for enhancement. Suppose that the pharaoh can choose between two options. He can seek to internalize enhancement or he can seek to externalize enhancement. He could seek to internalize enhancement by increasing the physical strength and endurance of his workers. He might dose them up with anabolic steroids, synthetic EPO, and respirocytes. The pharaoh might employ genetic engineers to introduce sequences of DNA associated with greater endurance or a stronger work ethic into the genomes of his workers. He might recruit some specialists in cybernetics to increase their strength by giving the workers robotic limbs. If successful, these measures would considerably speed up pyramid construction. Were enhanced workers to be available, the pharaoh might unhesitatingly use them to construct his pyramid. But a policy of internalizing the enhancement of pyramid building is a less efficient way to achieve the end of a more speedily built pyramid than is a policy of externalizing enhancement. A better focus than the physiques and attitudes of the workers would be machines that they might use to sculpt, haul, and place the massive stone blocks. Trucks and diggers are, from the perspective of a 2000 BCE pharaoh, futuristic technologies. But the fact that we've had them for a while seems to indicate that they are, from a purely technological perspective, less of a challenge than technologies that internalize the same functions. The pharaoh would have required a small fraction of the workforce required by his predecessors had he had access to early twenty-first-century construction equipment.

Those who seek an enhanced access to external goods by internalizing enhancement face a challenge that those who externalize do not. They face a problem of *integration*. They want to make enhancements that are part of human bodies and brains. The enhancement must be directly integrated with existing human physiology. Externalizers of enhancement require only that the enhancements be operable by humans. They cleverly avail themselves of efficiencies enabled by biological design. Our sense organs evolved to convey information from the external world, from the readouts of machines to the parts of our brains that make best use of it. We access the Google search engine by reading a computer screen bearing the information resulting from a search. This information is conveyed to the visual processing parts of our brains. Once processed, it is sent on to parts of the brain designed to extract meaning from it. Our limbs evolved to enable us to move things and throw things. They also enable the operations of gear-shifts and steering wheels. The policy of externalizing enhancement may be less satisfying from the perspective of a worker who would like to brag about how physically strong he is. But it's likely to lead to speedier pyramid construction.

I mean this to be a general claim about the relative instrumental value of internalizing as opposed to externalizing enhancement. Those who seek to internalize enhancement face a technological challenge that those who externalize enhancement avoid. There's no reason in principle why the information currently displayed on computer screens by the Google search engine could not be delivered directly to our brains by way of an interface that directly stimulates neurons. It is, at this point in our understanding of the brain, quite a challenge.[13]

Consider a particular proposal to increase the instrumental value of human brains. Ray Kurzweil is greatly impressed by the acceleration of our cognitive processing enabled by electronic upgrades. Our brains do much of their processing of information by way of electrochemical signals trans-mitted from neuron to neuron at a speed of 100 meters per second. The electronic signals of computers and electronic neuroprostheses travel at 300 million meters per second. According to Kurzweil, this means that an electronic functional counterpart of a biological brain would process infor-mation "thousands to millions of times faster than our naturally evolved systems."[14] The mere act of implanting electronic chips that are function-ally identical to the neurons and synapses that they replace would produce a quite dramatic increase in the speed with which our brains process infor-mation. We will have been radically enhanced, considerably increasing the instrumental value of our cognitive faculties.

We should compare this approach to enhancing instrumental value with approaches that involve externalizing this manner of improvement. Those who externalize enhancement have built a range of tools that enable the relatively slow human brain process to respond appropriately to fast evolving natural and artificial systems. Human pilots of space shuttles and controllers of nuclear power plants confront many choices that must be made more quickly than would be possible for unaided biological brains. They rely on computers to make these decisions. This seems to entail no loss of control or prestige. In the domain of physical strength we accept that some tasks cannot be performed by unaided humans. We accept that some objects are too heavy to be lifted by unaided humans. It's no affront to human dignity to build machines to lift them. It's no affront to human dignity to leave decisions that must be made very quickly to computer systems designed specifically to make those decisions. These computer systems would be designed and programmed by human beings. So we would retain control over how they make their decisions even if we don't make these decisions ourselves. Humans still fly space shuttles and control nuclear power plants. They leave certain decisions that must be made very quickly to their computer assistants.

In my earlier book *Humanity's End*, I explored a different reason for skepticism about the value of dramatically accelerating our thinking. It's possible that an acceleration such as that sought by Kurzweil might have dire consequences for our conscious experience. One difference between information-processing tasks of which we are conscious and information-processing tasks of which we are not conscious is the speed with which they occur. I remain unconscious of the work that my brain does turning information received by my retinas into an image of my son's face. I am conscious of the much slower mental processes that guide my response to his request for a chocolate cookie. Would the acceleration of the latter kinds of process shift them below some temporal threshold for conscious awareness? It's hard to say. Perhaps among the many factors relevant to consciousness is the speed of a given process relative to other processes of the *same* mind. A vast overall acceleration of my intelligence would preserve conscious awareness of my evaluation of my son's request by conserving the relative speeds of conscious and nonconscious processing. Alternatively, perhaps there's a physical law that denies consciousness to any processing of information that exceeds a certain speed. Consciousness is such a perplexing process that it is difficult to know whether there might be an absolute speed threshold. It is a potential threat to those who aspire to boost the instrumental value of human cognitive capacities. Such enhancements may come at a grievous cost in terms of intrinsically valuable conscious experiences.

We don't need to choose between a human option of capacities that enable experiences with which we can engage veridically but have comparatively low instrumental value and an option of radically enhancing capacities that do not permit experiences that we view as meaningful but have high instrumental value. The choice we must make is not so much about whether to accept or reject the goods enabled by the radical enhancement of our cognitive or physical capacities. Rather, it's about whether we should internalize or externalize radical enhancement. A policy of internalizing enhancement involves increasing the instrumental value of our mental and physical capacities. A policy of externalizing enhancement involves making technologies more instrumentally valuable in ways that do not require alterations to human beings. What's remarkable about the project of radically enhancing humans is not the idea of developing technologies that enable us to do remarkable things. It's the idea that these technologies should be integrated into human brains and bodies—that these technologies should be internalized. Internalizing them tends to reduce intrinsic value. The policy of externalizing enhancement is the idea that we can continue to receive the benefits of rapidly advancing technologies without being sucked up into them. In maximizing the instrumental value of our technologies, we can preserve the intrinsic value of our experiences and achievements.

This is not a particularly revolutionary proposal. We build machines to clear our driveways of snow. We don't feel the need to enhance the snow-shoveling aspects of the human body. We can build computers capable of calculating billions of places of π—we don't need to radically supplement the human mind to be able to do this. Directly integrating electronic devices into human brains is an additional demand that distracts from the task of calculating places of π.

Chapter 4 addresses the prospect of radically enhancing our cognitive faculties. I argue that the pursuit of understanding of the universe and our place in it does not warrant the radical enhancement of our cognitive faculties. I make use of the distinction between internalized and externalized enhancement to argue that a rejection of radical cognitive enhancement entails no limitation on what we can discover about the universe. It is unlikely to retard scientists' progress on the pressing problems of our age.

Is the Distinction between Internalizing and Externalizing Enhancement Philosophically Principled?

I have distinguished between two policies that have the potential to radically improve what we can do. We can internalize enhancement. This involves making our brains and bodies better. We can externalize enhancement. This involves making the external technologies that we use better.

Is the distinction between internalizing and externalizing enhancement philosophically principled? The possibility of such a distinction faces a challenge from the extended mind hypothesis.

Neil Levy offers a philosophically effective account of the extended mind hypothesis. He understands the thesis as holding that "the mind is not wholly contained within the skull, or even within the body, but instead spills out into the world." Levy continues that, according to the extended mind view the "set of mechanisms and resources with which we think" is not limited to "the internal resources made up of neurons and neurotransmitters. Instead, it includes the set of tools we have developed for ourselves—our calculators, our books, even our fingers when we use them to count—and the very environment itself, insofar as it supports cognition."[15] This attitude toward the mind and its boundaries is supported by a functionalist approach that categorizes mental states according to their roles in causing behavior. If an external process plays the same role in thinking as an internal process that we correctly recognize as mental, then the external process should be recognized as mental.

The extended mind hypothesis is, unsurprisingly, a hypothesis about the mind. However, it suggests a matching thesis about the body and enhancement directed at states of it. It seems to suggest the philosophical irrelevance of the distinction between internalizing or externalizing enhancement. It should make no difference to cognitive enhancement whether it is directed at neurons and neurotransmitters—internal resources—or calculators and computers—external resources. An external process might be correctly recognized as playing the same functional role as parts of our body. A forklift does the same job that we use our muscles to do. Designers of forklifts enhance human muscles just as surely as do those who modify the genes that make muscles so as to boost their bulk.

Andy Clark is an advocate of the extended mind hypothesis who places particular emphasis on the role of technology to extend our minds and bodies. According to Clark, we are cyborgs. By this Clark means that we are human-technology symbionts "thinking and reasoning systems whose minds and selves are spread across biological brains and non-biological circuitry."[16] The notion that there is a philosophical distinction between enhancement directed at human brains and bodies and enhancement directed at our machines may seem to assume a view about our identities that advocates of the extended mind hypothesis reject.

I do not wish to deny the utility of the extended mind approach, especially in psychological explanation. But the existence of a fruitful strategy of psychological explanation that dispenses with conventional boundaries does not rule out other philosophical purposes that respect the traditional boundaries of minds and bodies. For example, there are explanatory purposes in the human sciences for which it is appropriate to imagine away the boundaries between individuals. Macroeconomists study the properties of economies as wholes. The phenomena that interest them—price levels, inflation, different rates of economic growth—result from very complex interactions of large numbers of individuals and corporate entities in economies. Macroeconomic explanations do not seek to connect a change in aggregate demand to the actions of particular individuals. They do not seek to subdivide the group property of a rate of inflation, allocating parts of it to different individuals. Their lack of interest in the boundaries between individuals serves a specific theoretical purpose. Macroeconomists do not go so far as to suppose that these boundaries are not relevant to other explanatory purposes. They allow that other economists could be particularly interested in investigating the implications for the economy of the spending decisions of specific individuals. We can say the same about the explanatory approach of the extended mind. If Clark and Levy are right, there are important explanatory purposes that require forgetting the distinction between processes internal to the biological brain and process external to it. One can do long division in one's head or on a hand-held calculator. The states of the part of your brain dedicated to mathematics and the states of the calculator stand in the same broad causal relationship with the behavior you perform—for example, verbally reporting the answer. Physically attaching the calculator to your prefrontal cortex makes no difference. Psychology is a very important member of the collection of explanatory interests we have in human beings. The extended mind hypothesis certainly represents a significant revision of psychological explanation. But advocates of the extended mind have not shown that there are *no* explanatory or evaluative purposes that require a distinction between processes internal to the brain and body and processes external to them.

Consider one explanatory purpose for which the boundaries of the body are significant. Oncologists are interested in working out how a cancer that originates in one particular location spreads to other locations. There may be some contagious cancers, cancers capable of spreading from one organism to another. But the spread of cancer generally respects the conventionally recognized boundaries of bodies. If you are diagnosed with cancer, your oncologist might seek to discover other locations in your body the cancer might have spread from or to. Unless there is reason to believe that the cancer is one of the rare contagious forms of the disease, she will not subject the bodies of those with whom you have interacted to the same level of scrutiny.

Suppose that we accept Clark's view. We are cyborgs. In this chapter, I have explored two reasons for designing human-technology symbionts in a certain way. One reason points to an engineering consideration. When aerospace engineers design a new passenger jetliner they aim to keep certain parts of the plane well separated. For example, they seek to avoid having electrical wiring running through the plane's fuel tanks. It is likely to be difficult to functionally integrate new machine parts and biological systems. One engineering shortcut would be to build the former to be operable by way of movements of our body directed by our sense organs. Information is conveyed from nonbiological parts of our minds and identities into the biological parts of our minds and identities by way of senses specifically designed to convey information to the parts of our biological brains where it is best exploited. There's a good engineering reason to keep new parts of our cyborg-selves external to our biological brains and bodies.

There's a second reason that has less to do with engineering and more to do with the value we place on our experiences. Externalizing our cognitive and physical aids protects the intrinsic value of experiences. Suppose that the advocates of the extended mind are entirely correct about psychological explanation. An interest in explaining human behavior makes no distinction between accessing memories stored in your brain and accessing memories stored on your iPhone. This does not prevent us from placing different values on enhancements of our powers of memory achieved by modifying our brain tissue on the one hand and those achieved by procuring a smartphone with more room to store data.

If Clark and other defenders of the extended mind view are correct, then this is no challenge to the fact that we are cyborgs. It concerns what kinds of cyborgs we become. We should become the kinds of cyborgs whose new machine parts remain functionally isolated and physically separate from our biological parts. We should become the kinds of cyborgs who access the parts of our minds that contain the Google search engine by way of a keyboard or a touch screen rather than by way of a chip directly attached to our neurons and synapses.

4 The Threat to Human Identities from Too Much Enhancement

In chapter 3, I argued that radical enhancement is likely to replace very valuable experiences and achievements with less valuable experiences and achievements. This reduction in value is a consequence of an estrangement from the experiences and achievements of radically enhanced beings. In this chapter, I switch the focus of discussion to a problem that is effectively the inverse of that discussed in chapter 3. As radical enhancement provides us with less valuable experiences, it tends also to undermine the identities of those who undergo it. This undermining manifests as a threat to the connections of autobiographical memory that either explain human identity over time or explain a human being's sense of herself over time.

This chapter's investigation of the effects of radical enhancement on personal identity differs from many other philosophical discussions of the effects of certain kinds of change on human identities. In many discussions, the goal is to identify facts *constitutive* of our identities. These facts would be properly included on a list of necessary and sufficient conditions for human survival. If we decide that the application of a given technology disrupts a necessary condition for the preservation of human identity, then we know that the application of the technology necessarily ends a human individual's existence. We could not truly imagine humans surviving the disruption. Suppose that we decide that the continuance of an individual's distinctive psychology is necessary for the survival of a human individual. A technology that erases this distinctive psychology must necessarily bring to an end the existence of individuals to whom it is applied.

My concerns about the effects of radical enhancement on human identities do not conform to this pattern. I shall not claim that radical enhancement *necessarily* disrupts human identities. When we imagine a human being surviving radical enhancement we imagine something that is a genuine possibility. I make a more cautious claim. Radical enhancement is not logically or metaphysically incompatible with human survival. When you imagine yourself surviving a process of radical enhancement you are imagining something that could happen. We should nevertheless acknowledge that radical enhancement poses a threat to human identities. Compare: It's easy to suppose that you may survive skydiving without a parachute—you need only imagine the placement of a series of safety nets that break your fall. You should nevertheless recognize that skydiving without a parachute is imprudent even if your survival is easily imaginable. There's no logical necessity in death from skydiving without a parachute, but it should nevertheless be acknowledged as a probable outcome. By analogous reasoning, radical human enhancement is imprudent even if there is no logical or metaphysical incompatibility between it and survival. It is likely to end the existence of its human subjects.

Two Psychological Accounts of Personal Identity

The term "identity" appears in a variety of philosophical contexts. Psychological states feature in explanations of two different senses of identity. Some theories make them central to the *metaphysics* of human identities. A theory of the metaphysics of identity purports to answer questions about the boundaries of human identities. Suppose I want to know by virtue of which facts one of the seven billion people who will inhabit planet Earth in one year's time would be me. An account of the metaphysics of my identity should answer this question. It should indicate which, if any, of these future individuals is me. According to the psychological continuity theory of the metaphysics of human identities, one of these individuals would be identical to me if sufficiently many of my psychological states can be traced to him. If there is no such individual, then it's likely that I suffered some fatal misfortune between then and now.

A philosophical exposition of the psychological continuity account of the metaphysics of identity should begin with the views of the theory's best-known contemporary advocate, Derek Parfit.[1] Parfit analyzes personal identity over time—or more properly, survival over time—in terms of psychological connections. A psychological connection obtains when a psychological state, for example a memory, exists at one time and continues to exist at some later time. The connection obtains between person-stages—parts of persons that exist at specific times. For Parfit, I am essentially a collection of psychologically connected person-stages. A person-stage of me at forty is psychologically connected to a person-stage of me at forty-five by virtue of at least one psychological state in the earlier person-stage carrying forward to the later person-stage.

Some philosophers find the psychological continuity theory to be inadequate as an account of the metaphysics of human identities. Bernard Williams is among the best-known critics.[2] He presents thought experiments in which human individuals progressively lose their distinctive psychological states—their memories, beliefs, desires, and so on—yet intuitively survive. Williams concludes that we must look beyond continuities of psychological states to understand what underlies our survival over time.

There's another collection of questions about identity for which psychological states assume great significance. One can take an *evaluative approach to identity*. Those who take this approach are interested in what makes our continued existence meaningful or valuable. They ask what types of mishaps might erode or destroy the meaning or value we attach to our lives. It seems that an important precondition for such meaning or value is psychological – an awareness or sense of our selves. The evaluative approach to identity supports some complaints about neurological conditions that erase our memories. These conditions deprive a victim of any sense of who he or she is. As victims lose memories, they may cease to recognize loved ones. Parfit would describe the neurodegenerative condition as striking at the metaphysics of sufferers' identities. Those who agree with Williams and insist that one may determinately survive the loss of a sense of self may explain the tragedy of the situation in precisely these terms. The neurodegenerative condition is so tragic precisely because its sufferers fully survive but in a state deprived of a proper awareness of who they are and what matters to them.

A Threat to Identity from Life Extension

I am interested in the psychological bases of our identities for what they might tell us about dangers from too much human enhancement.

Walter Glannon draws on the psychological continuity theory of personal identity to question the value of radically extending human life spans. According to Glannon, "a substantial increase in longevity would be undesirable because it would undermine the psychological grounds for identity and prudential concern about the distant future."[3] Glannon offers for evaluation a possible human life span of 200 years. He proposes that facts about the psychological bases of our survival over time should give us little incentive to desire such a life span. Glannon's complaint about radical life extension can be expressed either in terms of the metaphysical or the evaluative approach to identity. Depending on one's view about what the psychological continuity theory explains, one can understand him as arguing that it's pointless for a human being to seek to extend her life span beyond 200 because facts about psychological connections show that no human being can live that long. Alternatively one can grant the possibility of extending human life spans to 200 years and far beyond, but insist that inspection of psychological connections show that this form of survival lacks much or all of what makes life worthwhile.

Glannon's skepticism about the prudence of life extension draws on a controversial aspect of Parfit's presentation of the psychological continuity theory. According to Parfit, focus on psychological connections reveals that our survival over time is not all or nothing. People have new experiences and acquire memories of them. They forget past experiences. They acquire new desires and cease to desire things that they formerly desired. There are likely to be greater numbers of psychological connections between "Nick Agar" person-stages closer in time than between "Nick Agar" person-stages more distant in time. Psychological connectedness comes by degree. This is a challenge to the commonsense view, which presents identity as all or nothing. Superman is either wholly identical to Clark Kent or not identical to him at all. It seems nonsensical to assert that the superhero is "just a bit" identical to the reporter. Psychological connectedness, by contrast, is not all or nothing. The reduction of connectedness that occurs with the advance of Alzheimer's disease may reduce connectedness to such an extent that very little of what matters in survival is transmitted from a pre-Alzheimer's person-stage to a post-Alzheimer's person-stage.

This is the root of Glannon's concern about radical life extension. He exposes a tension between radical life extension and the metaphysics of identity and survival. The connections between earlier and later stages of the 200-year life span "would be so weak that there would be no reason to care about the future selves who had these states."[4] Access to cures for all of the biological causes of aging should make no difference to your likelihood of living to 200.

Glannon's skepticism about radical life extension is bolstered by facts about how memories are processed and stored in the brain. Suppose that biological memory worked like a computer's memory. So long as the computer is not destroyed and no one pushes the delete button, then there's nothing that should prevent its information from being stored indefinitely. An owner determined to preserve his machine could resolve not to store new information on it. There would be no decline in informational survival. Glannon describes neurobiological mechanisms whose functioning seems to make our memory very different from computer memory. Healthy human brains undergo a process that involves purging existing psychological states to make space for new ones: "The action of a particular molecule, cyclic AMP response element binary protein (CREB), maintains equilibrium between learning and forgetting by modulating the formation-storage-retrieval process."[5] CREB comes in an "activator" form that permits the formation of new memories. It also comes in a "blocker" form that prevents the formation of new memories. The brain balances supply of these two forms. Too much blocker CREB leaves the brain unable to store crucial information. Too much activator CREB leads to "an overproduction and oversupply of memory, the mind becoming cluttered with memories of events that serve no purpose."[6] Glannon presents these two forms of CREB as "critical to the psychological unity between anticipation of the future and memory of the past, a unity critical to both personal identity and prudential concern about the future."[7] There's nothing to prevent clever bioengineers from seeking to modify this balance. Perhaps expanding the brain's capacity to store memories would enable the two forms of CREB to find a new balance between forgetting and remembering, one that permitted us to retain memories for longer. But this is a considerably more involved project than addressing the biological processes that are the proximal causes of aging.

To what extent should Glannon's skepticism deter would-be radical life extenders? John Harris retains his enthusiasm. Glannon offers a hypothesis about brain mechanisms and their relevance to the value of radically extended lives. Harris would like to see some willingness to test that hypothesis. He says: "The respectable scientific response to this would be not to insist that producing immortals is undesirable and therefore should not be done, but to say 'this is an interesting hypothesis, let's produce some immortals, see how their brains react and test the hypothesis.'"[8] Intending participants should view this experiment as having a quite considerable potential upside, while lacking a downside. If Glannon is wrong, the result would be "immortals who had the considerable advantage of very long life." If Glannon is right, then there are no immortals to receive this particular benefit, but no one suffers any harms predicated on living for too long. I think Harris may be right about individuals deciding whether to undergo a procedure proven to radically extend life. They may view themselves as having much to gain but nothing to lose. Glannon's skepticism is relevant to policy makers deciding to what extent, if any, they should support research into radical life extension. As we will see in chapter 6, life extension research is likely to be costly, both in terms of money spent and in terms of harms inflicted on human test subjects. Money spent in pursuit of the uncertain value of radical life extension, for example, cannot be directed toward the proven value of new hospitals and vaccination campaigns.

Radical Enhancement and Autobiographical Memory

Glannon presents facts about memory as casting doubt on some of the benefits claimed for radical life extension. In what follows, I argue that radical enhancement of our cognitive or physical powers poses a direct threat to the psychological connections that explain a person's sense of herself over time. This threat is to a psychological state that is of particular significance to our self-awareness—autobiographical memory. A person's autobiographical memories refer to events or facts about his or her life. It's a feature of many of the most emotionally resonant autobiographical memories that they are rich in sensory-perceptual information often seemingly recorded from a first-person perspective.

Autobiographical memories feature prominently in presentations of the psychological continuity theory. The earliest versions of the psychological continuity theory, for example that due to the seventeenth-century English philosopher John Locke, were memory theories. Continuities of autobiographical memories were held to define the boundaries of persons. Five minutes ago there was a person typing on this laptop who was able to summon up autobiographical imagery of a first day at school. Now there's a person typing on the computer who's able to summon up essentially the same imagery of that event experienced from the same first-person perspective. This is but one of the many memory connections that bind person-stages into the person who is me. Modern versions of the psychological continuity theory add other psychological states to autobiographical memories. My survival from one minute to the next, one hour to the next, and one year to the next is partly a consequence of the persistence of a vast collection of memories, beliefs, desires, fears, lusts, and so on. But autobiographical memory seems nevertheless to be of the utmost importance. Discussion and debate with others may lead us to substitute some of their beliefs and desires for our own. A friend recounting her vacation may give us new beliefs about distant lands and new desires for exotic foods. It would be absurd to cite concerns about the integrity of one's identity as reasons for not asking people about their vacations. We would view such conversations with greater suspicion if they caused substitutions of autobiographical memory—if we found that our former autobiographical memories of the time she was away had been replaced by autobiographical memories of her trip.

The disruptions of psychological connections described by Glannon are presented as the consequences of normal neurological functioning. Sustaining identity over two hundred years and beyond just doesn't seem part of the design specification of a normally functioning human memory. In what follows, I present the radical enhancement of our cognitive and physical capacities as a disruptive influence on psychological connectedness. Too much enhancement actively interferes with continuities of autobiographical memory. Glannon argues that there is little to gain from radical life extension. Harris counters that there's likely to be little to lose. In what follows, I argue that those who undertake to radically enhance their psychological or physical capacities have much to lose. They shouldn't take the attitude Harris recommends for radical life extension, which is basically to give it a go and see what happens.

The threat to identity from radical enhancement takes different forms on metaphysical or evaluative approaches to the role of autobiographical memory. In neither case is it a necessary consequence of radical enhancement. I allow that it is both conceptually and logically possible for radical enhancement to leave sufficient autobiographical memories in place for us to survive the process with an intact sense of who we are. There's certainly no logical requirement that it will erase our autobiographical memories. But this is a somewhat likely outcome. Radical enhancement should be viewed as reckless—not a procedure that you should view as likely to be in your best interests. It's a bit like choosing to vacation in a war zone. There's no law of logic that directs that all Westerners who vacation in the Afghanistan's Helmand province in 2012 must die. But it is nevertheless a reckless choice of vacation destination.

How Does Autobiographical Memory Work?

How would radical enhancement threaten autobiographical memory? Consider the threat posed to our identities by a condition that has the opposite effect on our cognitive capacities—the neurodegenerative disorder Alzheimer's disease. Alzheimer's is the most common cause of dementia. It affects a wide range of cognitive functions. Many of these bear on autobiographical memory, either directly or indirectly.

Your autobiographical memories are not simple recordings of experiences and events relevant to your life. The memory of a first day at school is not like a video recording of the event filmed from your first-person perspective that runs with perfect fidelity whenever the "play" button is pressed. If this model of memory were correct, the destruction of this significant autobiographical memory by Alzheimer's would require the deletion of a file that specifically encoded that event.

A more sophisticated model of memory leaves many more opportunities for Alzheimer's to have deleterious effects. According to this more sophisticated model, our brain does not store complete recordings of events.[9] What are recorded are schemata encoding salient information about remembered events. The act of remembering fills out a schema. This filling out draws on a wide range of knowledge and cognitive abilities.

In Alzheimer's disease, neurological damage affects this reconstructive aspect of autobiographical memory. Late in the disease's progression, people lose the capacity to recognize loved ones. This loss affects a wide range of autobiographical memories involving those loved ones. When one loses the capacity to recognize one's husband, one is also likely to lose or to have more restricted access to the autobiographical memories involving the husband—the memory of the first kiss, the memory of the wedding day, and so on. The destruction of the parts of the brain that store knowledge of activities has a negative effect on autobiographical memories involving those capacities. Late in the progression of his Alzheimer's disease, Ronald

Reagan is said to have forgotten that he was ever president of the United States—an office that he held for eight years. This loss is unlikely to have occurred by way of the disease destroying all of the very many individual autobiographical memory files of things that he did and things that happened to him while president. Rather, the loss occurred as a consequence of the loss of knowledge of a range of activities—planning of election campaigns, grasping political information conveyed by aides, making historic speeches, and telling hokey jokes—that were characteristic of Reagan's holding of that office. The loss of these capacities is likely to have blocked the reconstructive processes required for Reagan to recall his experiences as president.

Advocates of the psychological continuity theory as an account of the metaphysics of human identities may see Alzheimer's as effecting the gradual destruction of our identities. Alzheimer's can have fatal outcomes even as it leaves a living breathing human body. The Alzheimer's patient remembers nothing of his past. His diseased brain sustains few of the beliefs and desires that it formerly did. There is a radical diminution of the psychological connectedness that accompanies and, according to advocates of the psychological continuity theory as an explanation of the metaphysics of identity, explains survival. Opponents of the psychological continuity theory as an account of the metaphysics of identity may resist this conclusion. According to them, Alzheimer's disease is such a misfortune precisely because the original person's identity remains intact. Relatives would be wrong to view the person in the body of their uncle as just another confused rest-home resident. By wiping out a person's autobiographical memories, it destroys much of what matters in survival.

Radical cognitive enhancement has effects on human identities that resemble those of radical cognitive decline. It tends to eliminate autobiographical memories. The mechanism differs from that in Alzheimer's disease. Brain tissue is not destroyed. Or if it is, the destruction occurs to facilitate its replacement by electronic chips that perform its functions better. But nevertheless, as our brains undergo enhancement, the reconstructive capacities central to the maintenance of autobiographical memories are affected.

You are less likely to retain autobiographical memories of your past if enhancement makes the events that they refer to less remarkable and therefore less memorable. This is because they are now evaluated relative to standards that make them seem less remarkable. A marathon performance which is memorable because it pushed you close to human limits is entirely unremarkable and unmemorable relative to radically enhanced standards. Infusing your body with nanotech hemoglobin enables you to run faster marathons. In doing so, it makes your past marathon times less memorable. Your past marathon efforts are remarkable and therefore memorable when judged relative to human norms. You are right to view your 3 hour, 30 minutes time as very impressive. It might feature on a list of your life's most significant achievements. Consider what happens once you undergo radical athletic enhancement. Your nanotech enhancement means that you now evaluate it relative to different standards—those of a radically enhanced athlete. Your formerly impressive achievement becomes merely feeble.

Suppose you've written a book defending radical enhancement of which you are very proud. Arguments of sufficient quality to convince humans are unlikely to satisfy the elevated intellectual standards of judges with radically enhanced cognitive powers. Viewed from the perspective of your future radically enhanced intellect, it's an extremely poor piece of work. Ray Kurzweil is a fearsomely intelligent man. There is much that he has done of which he is justifiably proud. He anticipates cognitive enhancement that will make these achievements seem merely feeble.

Note that this loss of significance occurs with the dramatic increase in our powers. It's not something that occurs with incremental improvements. Suppose that as a consequence of intensive training you complete your second marathon in 3 hours, 20 minutes. This is more objectively impressive than your earlier time. But your new athletic powers do not bring with them evaluative standards that make the earlier effort seem trivial and without significance. Exercises that enhance your cognitive powers to a moderate extent may lead you to acknowledge a flaw in your earlier defense of radical enhancement. But you probably do not view your earlier arguments as merely feeble.

Transformative change brings about a shift in evaluative frameworks. What we greatly value before we value less after. What we place little value on before we value greatly after. Your past experiences become less relevant because the past events to which they refer are properly viewed as less noteworthy. Autobiographical memories become less likely to be retained.

Consider the implications for the transformative change of body-snatching for your identity. If Nozick's argument is correct, then you should survive the process of being snatched. But an interest in preserving psychological connections should lead to doubts about the long-term survival prospects of humans who become pod-aliens. They are likely to go on to have experiences that are not only different, but very different. The irrelevance of human memories to current pod-alien existences is likely to lead to a progressive forgetting.

Some works of science fiction attempt to engage with radical human enhancement. Among the most interesting and considered of these comes from the Scottish writer Iain M. Banks. A series of Banks's novels features a technologically advanced interstellar society called the Culture. The citizens of the Culture are highly diverse—they include many species of humanoids, a wide variety of nonhumanoid biological beings, and artificially intelligent machines. Transformative changes are commonplace events in Banks's fictional universe. Culture citizens are free to switch gender or species pretty much at will. One of the transformations on which Banks offers some detail happens to Djan Seriy Anaplian, a character in the 2008 novel *Matter*. Anaplian was born of the Sarl, a humanoid species that does not belong to the Culture. Banks presents the Sarl as an unusually warlike people who have attained a technological level that is the approximate equivalent of Europe's in the sixteenth century. Anaplian comes to join the Culture, whereupon she receives a variety of quite staggering enhancements. Modifications of her bones and muscles make her significantly stronger and faster. Her senses are dramatically extended. She senses radio waves. She is able to operate nearby machines merely by wishing to. She switches off pain and fatigue at will. A device called a neural lace implanted into and throughout her biological brain enables some of these new abilities. Some of the new functions are powered by an antimatter battery placed inside of her head.

What are the consequences of these changes for Anaplian's autobiographical memory? What should we expect that Anaplian, the human–machine hybrid Culture agent, will remember about the life of Anaplian of the barely postmedieval Sarl? Banks depicts Anaplian as having an ambivalent attitude toward the past. It's obvious why she might think this way. The Sarl are an obdurately patriarchal people. Though royal born, Anaplian's prospects are severely limited. She is still young when she joins the Culture. It does seem understandable that she might want to scrap her past existence. Such a transformation makes sense to the extent that we place diminished value on the existences and experiences we had before we undergo transformative change. Perhaps hers is a case in which complete alienation from her past life makes sense. Those who value their unenhanced experiences are more likely to view this alienation as an unwanted by-product of radical enhancement.

In the final section of this chapter, I argue that the age at which one undergoes radical enhancement makes a difference. We accept and indeed value the alienation from our childhood brought by the process of cognitive and physical enhancement we call growing up. We do not think this way about alienation from our adult commitments and projects.

An Asymmetry in Our Attitudes toward Past and Future

The previous paragraphs may seem to overlook a major difference between Alzheimer's disease and cognitive enhancement. Alzheimer's disease deprives one of the ability to recall significant events in one's past. This is most certainly not the case with radical enhancement. Here the barrier to autobiographical memory is the creation of new evaluative standards that make many of the significant events in one's life unworthy of being remembered. But one could presumably remember these events if one truly wanted to. Radically enhanced individuals could presumably seek to retain the evaluative standards that would facilitate remembering events in a life previous to enhancement.

Threats to our identities are consequences of the different agendas that too much enhancement tends to create. Too much enhancement creates a difference in agenda between the pre-enhancement selves and the postenhancement selves. Our postenhancement selves are likely to act in ways that pay little or no heed to maintaining connections with our pre-enhancement selves. This is something that those contemplating enhancement should consider. If you give another agent power over your life you should ensure that she takes your interests seriously. Too much enhancement is one way to surrender power over key interests to an agent who you have good reason to believe will be indifferent to them. In this case, that agent is a future self.

The suggestion that our postenhancement selves will be indifferent to the possibility of connection with our pre-enhancement selves does not suppose any malign intent. Rather, it's a consequence of a feature of prudential rationality that directs us to worry more about connections with our futures than with our pasts. The asymmetry in our attitudes toward the past and future makes evolutionary sense. Humans are designed to worry about the future because their actions can have fitness-boosting effects on the future. It's not possible to boost fitness by altering the past.[10]

For an example of this difference in attitude toward the past and future, consider two different life extension proposals:

Life extension option 1   You learn that you will suffer from Alzheimer's disease at some point in your future. The disease will progressively eliminate your autobiographical memories. You present at a clinic that administers a therapy that will retard the disease's progress by precisely one year.

Suppose we assume that the psychological theory captures the metaphysics of identity. The therapy extends the length of your life. This *prospective* extension of your life by one year seems truly worth having. Compare it with what we might call *retroactive* life extension. It involves a process known as rebirthing. According to its advocates and supposed exponents, rebirthing involves remembering the experience of being born. They claim that this will help us to deal with negative thought patterns that result from the trauma of birth. This is all highly dubious. But suppose, for the sake of argument, that rebirthing really did work. It might be a way not only to deal with psychological problems but also to extend your life span.

Life extension option 2   You present yourself at a rebirthing clinic. A variety of techniques including hypnosis and deep brain stimulation enable you to recall both the event of your birth and events of your early infancy. These therapies give you autobiographical memories that extend one year earlier than your earliest such memories prior to the therapy.

The psychological continuity account traces your existence back to the earliest autobiographical memories and intentional states connected to your current person-stage. The earliest experiences produced in your brain are, according to this view of identity, unlikely to be yours. According to current estimates, the fetus acquires the capacity for consciousness at some point early in the third trimester of its development. This is the stage in the brain's development at which the thalamocortical complex, a region important for consciousness, begins to come properly online. Any experiences produced by the brain early in the third trimester are, according to the psychological continuity theory, unlikely to belong to you, or indeed to anyone. They remain in isolation from any collections of time-slices of experiences that we would properly call a person.

If your brain did produce experiences that predate your existence, then rebirthing might be a method of life extension. Suppose that rebirthing is a way to establish connectedness between your present self and selves housed in your biological brain just before you were born. If the psychological continuity theory is a correct account of the metaphysics of identity, then it would be correct to say that your existence now extends further back into the past than it formerly did. You will have undergone retroactive life extension. There need not be anything metaphysically spooky about this establishment of connections backward in time. It requires that the experience of being born has left traces in your body or mind. In the normal course of events, you are disconnected from these traces—they never generate veridical memories. Presumably the rebirthing process would establish connections to the past by turning these traces into veridical autobiographical memories of the experience of being born.

The previous paragraphs assume that the psychological continuity theory explains the metaphysics of our identities. Alternative views of the metaphysics of human identities would not allow rebirthing to be presented as a form of retroactive life extension. But they can be seen as extending a certain interest we have in our past. The creation in us of autobiographical memories of early infancy extends our senses of ourselves. We become aware of our lives as they were at birth and immediately after birth.

Exclude, for the purposes of this example, any therapeutic benefits that might follow from rebirthing. It seems rational to prefer any amount of prospective life extension to any amount of retroactive life extension. Consider the metaphysical account of identity. Any additional months gained by rebirthing are months that you've already lived. The difference between option 1 and option 2 is the difference between a windfall that you've already enjoyed and one that you will enjoy sometime in the future. Our prudential concern in respect of extending our lives and receiving windfalls looks to the future and not to the past.

The preceding paragraphs have focused on psychological approaches to the metaphysics of human identities. Analogous implications hold for the evaluative approach to identity. Some age-related diseases destroy the autobiographical memories that account for our awareness of ourselves. Suppose that the psychological continuity theory does not account for the metaphysics of our identities. Autobiographical memories should neverthe-less be acknowledged as central to our sense of ourselves. Suppose option 1 protected you against Alzheimer's, extending your future self-awareness by one year. Option 2 retroactively extends your sense of self by one year. You are now aware of the life that you led immediately following your birth. Option 1 seems prudentially preferable. The experiences added to your sense of self by option 2 are experiences you have already had.

The Tension between Enhancement and Survival

Suppose that you are an unenhanced human contemplating a quite con-siderable enhancement of your cognitive abilities. You have an interest in enhancing your powers. You also have an interest in surviving the enhance-ment procedure in a way that preserves what matters to you about yourself. The metaphysical and evaluative presentations of the psychological con-tinuity theory point to two identity-related ways in which enhancement could be a disappointment. Both of these involve a threat to autobiographi-cal memory. Both suppose that the enhancement procedure succeeds in enhancing—the being who emerges from it is superior to the being who was the procedure's subject. According to the metaphysical interest in your identity, the enhancement procedure could fail if it terminates your existence, replacing you with a numerically distinct being better than you in some significant respect. According to the evaluative interest in your identity, the enhancement procedure may not terminate your existence. It nevertheless makes your existence poorer by predictably failing to preserve some or all of what matters in survival. It tends to eliminate memories of events that you view as very significant to your life.

Suppose that you self-administer an enhancement technology that considerably boosts your cognitive abilities. This change does not in itself affect other psychological traits. It does not directly delete or modify auto-biographical memories. It does not directly modify any significant beliefs or desires. Your pre-enhancement self has a strong interest in the mainte-nance of these psychological connections. Your postenhancement self is likely to have a weaker interest in maintaining those particular connec-tions. They connect to a past self that lacks many of the capacities that are now central to your life. There is an asymmetry that empowers our later enhanced selves at the expense of our earlier unenhanced selves. For the postenhancement self, it's all onward and upward. The postenhancement self has a much stronger interest in its connections with future experiences than in any connection with past experiences.

This seems to describe the predicament of Djan Seriy Anaplian, formerly of the Sarl and now of the Culture. She has undergone radical cognitive and physical enhancement. These changes differ from those produced by Alzheimer's disease. They do not prevent her from maintaining an appropriate connection with the significant experiences, beliefs, and achievements of her Sarl-self. She could deliberately retain standards that present these experiences, beliefs, and achievements as valuable. These standards would ill serve her as an agent of the fabulously technologically advanced Culture. Furthermore, an asymmetry between attitudes toward our pasts and futures that is part of prudential rationality directs that she should be comparatively indifferent about maintaining links with past experiences, beliefs, and achievements but very concerned to ensure those links with future experiences, beliefs, and achievements. Anaplian's Sarl memories are simply not relevant to the life she currently leads. They are likely to be lost together with the standards that make them memorable.

On the view that presents autobiographical memory as significant to the metaphysics of our identities, Culture enhancement is a process that is likely to terminate our existences. On the view that grants autobiographical memory an evaluative significance, those anticipating Culture enhancement may survive, but will be deprived of much of what they currently view as significant.

The Analogy with Childhood

I think we should take the same approach to radical enhancement that we currently take to cyberconversion. Is it possible that I'm exaggerating this problem of the loss of significance of our past experiences and achievements? Radical intellectual and physical enhancement is something that almost all of us have direct experience of. As newborns, we know next to nothing and are capable only of feeding, sleeping, and complaining. We undergo a dramatic process of enhancement to acquire adult mental and physical abilities. As adults, we retain relatively few memories of childhood. Psychologists offer a variety of explanations for the phenomenon of infantile amnesia—the seemingly complete absence of memories in adults of the first three to four years of our lives. According to some accounts, young children lack a fully developed concept of self to pin experiences to. Deprived of any mooring to a self-concept, experiences are soon forgotten. Perhaps there's an additional mechanism of forgetting. As we mature, we undergo cognitive, emotional, and physical enhancement. We apply new standards that make past experiences seem less interesting and therefore less memorable.

The example of the enhancement that we undergo as we mature seems to lead to the reverse conclusion from the one I have drawn about radical enhancement. For example, Nick Bostrom and Toby Ord make the point that it can be good for children to grow up, and that "it might be good for adults to continue to grow intellectually even if they eventually develop into rather different kinds of persons."[11] Growing up is a time of quite dramatic intellectual and physical transformation. We do have to deal with growing pains. But it's (almost always) worth it. Growing up certainly seems preferable to spending a normal human life span in childhood. If truly dramatic intellectual and physical improvement is good for a four-month-old, then why shouldn't it be for a forty-year-old?

There is something a bit tragic about growing up. As we mature, we lose the capacity to experience certain achievements as meaningful. For example, the experience of counting all the way up to a hundred for the very first time is meaningful for a child. As an adult, one can appreciate that this is quite an accomplishment for someone with a pretty shaky mastery of number concepts. But we don't view it as meaningful. Counting to a hundred is a remarkable achievement for a five-year-old. It's a tedious and laborious undertaking for a forty-year-old with a more secure grasp of number concepts—something we undertake only when forced to arrive at an accurate tally of a collection of objects. The experience does not warrant preservation. Many of our childhood achievements are largely lost to us. We accept this loss as part of the process of growing up.

I will make two points about Bostrom and Ord's comparison of radical enhancement as an adult with the seeming radical enhancement we all undergo as we mature into adults. First, discontinuities between postenhancement and pre-enhancement experiences are likely to be more pronounced than discontinuities between adult and childhood experiences. Second, although it is generally a good thing for a child to grow up, it does not follow that is it generally good for an adult to go through a similarly dramatic transformation. Bostrom and Ord's comparison points to a relative similarity between childhood and adulthood on the one hand and adulthood and a radically enhanced existence on the other. However, there are differences between human childhood and adulthood that Bostrom and Ord's comparison necessarily omits. These differences make it right for most people to reject a suggestion that they treat their adulthood as a second childhood.

Making sense of this comparison requires an analysis of the concepts of adulthood and childhood. What does it mean to be an adult? Since our focus is on humans, we might seek to give a list of the mental and physical traits that human adults tend to possess. This list might include information about the heights that humans tend to grow to, the cognitive abilities that humans tend to acquire as they mature, and so on. This chauvinistic approach to adulthood would render the concept valueless when applied to members of other species. For example, adult extraterrestrials are likely to lack many of the physical characteristics of adult humans. Here, I define adulthood in terms of the role it occupies in the life of a certain kind of being. Only certain kinds of beings can become properly adult, according to this understanding. For a species to have a stage properly recognized as adulthood, it must be capable of forming complex, all-encompassing desires about the direction its life should take. We can mark the onset of adulthood with the achievement of that capacity. Many species lack a developmental stage that we might define as adulthood. Earthworms reach biological maturity. But they are not the kinds of beings that can form life plans or entertain complex desires about the direction that their lives should take. Facts about the tendencies of mature earthworms come no closer to answering the question about what mattered to the earthworm than do any facts about juvenile earthworms.

Adults may change their minds about many things. They may mistakenly desire certain things and come to recognize this error. But adulthood is, nevertheless, a stage characterized by the physical, cognitive, and emotional resources to arrive at final and decisive plans about one's life. There should be no higher authority in respect of questions about an individual's own basic values and interests.

Adulthood is not the final stage in many human lives. Some undergo cognitive decline and become demented—they lose the cognitive abilities characteristic of adulthood. Consider our decisions about how to treat people in this state. It seems right to treat previous adult desires as authoritative in respect of choices that concern a demented person.[12] We rightly privilege adult interests even when adulthood may take up a comparatively short part of an individual's life. In answer to the question "What did that person stand for?" it is appropriate to cite adult projects, even if the person dies early in that stage of their life.

This approach to adulthood suggests a matching view of childhood. This approach does not identify childhood with a list of characteristics typical of human children. Rather, childhood is defined in relation to adulthood— the state of full responsibility over the direction one's life takes. Childhood is a stage that anticipates and prepares for the later stage of adulthood. On this approach, the mere fact that a desire about the direction a life should take is expressed by a child necessarily deprives it of authority. This is so even if the reason may be preserved into adulthood whereupon it acquires the necessary authority. Some lives do follow paths mapped out in childhood. Some little girls who desire to become nuns grow into women who choose to enter nunneries. Some little boys who fantasize about joining the army enlist as adults. In some cases, the reasons may be preserved. The little girl's desire for a life of piety free of the distraction of boys is expressed in more sophisticated language by the adult she becomes, but it's essentially the same reason. When expressed by a child, such reasons necessarily lack justificatory force that they acquire in adulthood.

This isn't to say that there are no foolish adult desires. You should stop a friend from deciding how to invest the entirety of her retirement savings while inebriated, because you doubt that she's in a position to make choices truly sensitive to her interests. Parents have a different justification for preventing their children from acting on life-altering desires. Your daughter's desire to become a nun may not suffer any of the rational deficiencies of the inebriated investor's choice. It may reflect a truly considered vision of the direction her life should take. The desire may result from calm, rational reflection on her current values and how best to realize them. It does not matter how rational a child's desires are. Some children's desires are better informed than those that they form as adults. They are likely to be responsive to a parent's possibly well-informed beliefs about the world. In many cases, the passage to adulthood replaces a somewhat accurate parental worldview with a more erroneous one shaped by an individual's own reflections and experiences.

Parents rightly prevent their children from acting on desires that may comprise later adult interests. A parent refuses to comply with her eight-year-old son's genuine desire to join the army, no matter how keenly the desire is felt and how well informed by relevant facts about soldiering and geopolitics it is. To permit such choices is to misunderstand the proper relationship between childhood and its interests and adulthood and its interests.

To anticipate radical enhancement is to put yourself in a position in which you should defer to the future higher interests of your radically enhanced self. It is to grant your current desires only a very limited authority over your life. Most of us would like to think that we could muster some kind of answer to the question "What do you stand for?" Those anticipating radical enhancement should really answer that question with "I don't know yet." Confident answers to these questions are like children's keenly felt desires to join nunneries or the army. They should be treated as dubious predictions of the desires of future, more enlightened selves.

One of the duties of human parents is to prevent children from doing things that are likely to conflict with their adult interests. Those among us now hoping for radical enhancement lack radically enhanced guardians to police their desires. But they should nevertheless acknowledge that their current desires and plans should be assessed not in terms of how they contribute toward projects that they currently deem to be worthwhile, but in terms of how they contribute toward unknown and indeed unknowable projects endorsed by radically enhanced future selves.

Compare this threat to our current adult selves with the famous critique of utilitarianism presented by Bernard Williams.[13] Utilitarianism is a consequentialist moral theory. It requires us to choose always to act in a way that produces the most good, overall. This is good impersonally conceived. What matters is the fact that good or bad consequences are produced, not who experiences them. Williams argues that utilitarianism's commitment to maximizing impersonal good poses a threat to our integrity. It alienates us from our projects. We are, according to Williams, debarred from writing poetry purely for the love of it. If we write poetry, we should do so only if doing so is the best way to contribute toward the overall good.

I think that the prospect of future radical enhancement has a more corrosive effect on our current commitments and projects than that hypothesized by Williams to arise in respect of moral consequentialism. At least agents who respect the authority of consequentialism over their current projects have a sense of the region within which they are at liberty to select their projects. They proceed with some manner of understanding of the demands that consequentialist moral theory makes of them. Those who expect to sign up for radical enhancement can have no such confidence in respect of the future projects that their more rational selves may form. The most keenly experienced commitments of children are often those of which their adult guardians should be most suspicious. We lack radically enhanced guardians to ensure that what we do in this intermediate childhood does not preempt or prejudice the commitments and projects of our future posthuman adulthoods.

This approach to defining childhood and adulthood enables us to identify two patterns that our lives could take:

The Human Pattern

Childhood → Adulthood

The Posthuman Pattern

Childhood → Second Childhood (Human Adulthood) → Posthuman Adulthood

In what follows, I argue for the superiority of the first, human pattern. I shall offer species-relative reasons. This is simply to say that this pattern is the best for members of the human species. Radically enhanced posthumans may rightly view this life as tragically shortened. A posthuman who, through misfortune, lived a life that followed the human pattern would be viewed as never reaching a stage at which she could have any truly authoritative sense of what her life was about. The ideal plan for a posthuman life is the posthuman pattern. Humans are no more required to echo this view than they are to echo the views of the bodysnatching pod-people.

Why Radical Enhancement Is More Psychologically Disruptive Than Growing Up

Human adulthood grows endogenously out of human childhood. Childhood is, by design, preparation for adulthood. The experiences of childhood prepare for adulthood. They are by-products of education or training involved in developing adult abilities. This relevance makes them memorable. When Bobby Fischer looked back on the chess games he played as a six-year-old, he is likely to have remembered play that was truly appalling when evaluated relative to his adult chess standards. But he also has insight into the training process that gave rise to his mature chess skills.

The conditions that drive radical enhancement are exogenous. Radical enhancements of our capacities emerge through the application of a variety of technologies to our brains and bodies. This means that past experiences are not precursors to enhanced experiences in the same way as childhood experiences are precursors to adult experiences. Enhancement technologies are likely to generate new capacities without any requirement for effortful learning or laborious training. Suppose you were to enhance your mathematical skills by way of attaching a math neuroprosthesis to your brain. The math neuroprosthesis radically enhances your mathematical skills by replacing or making redundant the parts of your brain that had been trained by years of schooling and practice at adding, subtracting, and otherwise manipulating numbers. It stands in a completely different relationship to your earlier experiences of adding and subtracting numbers than did the parts of your biological brain that it replaces. Your previous experiences learning mathematics are irrelevant to the exercise of your postenhancement mathematical skills. Other means of radical enhancement may not produce such a sharp discontinuity between pre- and postenhancement exercises of a capacity. Suppose one enhances mathematical skills by modifying genes that influence mathematical skills. This genetic enhancement may not make past experience with numbers completely irrelevant to postenhancement mathematical talents. But it does make them significantly less relevant. It is false to think of your postenhancement mathematical abilities as emerging out of exercise of your pre-enhancement abilities.

This seemingly great advantage of enhancement technologies, its propensity to take the blood, sweat, and tears out of skill acquisition, has a downside in respect of the sense we have of our selves. The self-understandings of radically enhanced people will track to these technologies more than to any formative activities or experiences of their pre-enhancement selves. The continuities of memories that connect pre- and postenhancement person-stages into a single person are systematically undermined.

The Regress Problem: The Tragedy of Unending Enhancement

There's an additional problem with the argument for enhancement beyond human adulthood. A consequence of accepting that it is prudentially good to enhance is that one could be subject to a prudential requirement to continue enhancing *ad infinitum*. Here is a third possible pattern our lives could take. Compare this with the human and posthuman patterns:

The Singularitarian Pattern

Childhood → Second Childhood (Human Adulthood) → Third Childhood (Posthuman Adulthood) → Fourth Childhood → ... nth Childhood → ...

The Singularitarian pattern takes its name from the technological Singularity—a predicted consequence of exponential improvements of technology. Kurzweil describes the Singularity as "a future period during which the pace of technological change will be so rapid, its impact so deep, that human life will be irreversibly transformed."[14] Here I am interested in the consequences of ongoing improvements of the technologies that enhance human capacities. Kurzweil's law of accelerating returns will give us increasingly powerful enhancement technologies. The law applies to technologies external to the human body and mind. But it applies also to technologies that operate internally.

If the law of accelerating returns operates freely within the boundaries of our brains and bodies, it has truly dramatic consequences. We will trade in computationally clunky, disease-prone neurons and synapses for electronic circuits. A willingness to renounce biology for electronics enables an increasingly miraculous, and from our present biological human standpoint, barely comprehensible series of transformations. Machine-human intelligences who have forsaken neurons and synapses won't hesitate to trade in electronic circuits for even more powerful media of thought. Kurzweil predicts that by the end of the twenty-first century, "the nonbiological portion of our intelligence will be trillions of trillions of times more powerful than unaided human intelligence."[15] This occurs through an exploitation of the computational potential of matter and energy. Kurzweil envisages our minds colonizing ever-increasing quantities of the universe's inanimate matter and energy. He offers a clear statement of where the law of accelerating returns is taking us and the universe: "Ultimately, the entire universe will become saturated with our intelligence. This is the destiny of the universe."[16]

Does this possible future leave room for a stage that we might identify as adulthood—a developmental stage that does not defer to the superior rationality of a later stage? The issue of an endpoint that might correspond with a post-Singularity adulthood depends, in part, on claims about the finitude or infinitude of the universe. The question of the finitude of the universe is controversial. Suppose that the universe is finite. We might achieve post-Singularity adulthood when our intelligence saturates the entire universe. If we have optimized the computational potential of all of the matter in the universe then there will be no further scope for cognitive enhancement. This colossal intellect might feel displeasure at its current cognitive limitations if it were to assent to the claim that were there to be additional scope in the universe for cognitive improvement then it would choose to enhance further. It will feel frustrated at the limits imposed by the universe on its cognitive growth.

Suppose, however, that the universe is infinite. There is an unending quantity of matter and energy to turn into the stuff of thought. There could be no such thing as a post-Singularity adulthood, a stage in our cognitive development that does not defer to the superior wisdom of a later stage in cognitive development. Ongoing cognitive enhancement will make our minds vastly capacious. But, in spite of this enormous capacity, there seems room in Singularitarian minds for one sincerely and deeply held project—namely, the project of cognitive enhancement. It is the only motivational constant that can be traced throughout a life that follows the Singularitarian pattern.

Are there ways to defend an adult stage? How could one possibly justify stopping when the option of further cognitive enhancement exists? Suppose we are confronted with the option of radical enhancement now. We reason that a radically enhanced existence must be superior to our present existence. Once we radically enhance, we find ourselves confronting the prospect of an additional quantum leap. What reason could we give for rejecting this new leap in power that could not also be given for rejecting the first leap? Should those who desire to reach an adult stage in development, a stage at which they could be deemed to have full authority over the direction of their lives, randomly nominate some stage as adulthood? Maybe they would be deemed to have reached adulthood after a certain number of applications of enhancement technologies.

Perhaps there is something substantive to be said in favor of a human pattern in which a phase of quite significant cognitive and physical enhancement follows a period of childhood that cannot be said of the post-human or Singularitarian patterns. Children lack the capacity for fully considered life plans. This is an absolute claim about human children rather than a claim about the cognitive and physical abilities of children relative to adults. Children lack any awareness of how their many current commitments might combine to promote or hinder a plan for their entire life. This simply isn't the way they think about their lives. Something that we are right to dignify with the label "life plan"—a considered statement of the best life for you and what really matters in life—arrives with human adulthood. Radical enhancement could replace this human life plan with something altogether more complex. But it's at this stage that there is something prudentially weighty to set against radical enhancement. Adult humans can reject the degree of enhancement that threatens to trivialize these plans. On this account, the transition from human child to human adult that is the central feature of the human pattern differs from the transition from second childhood to posthuman adulthood that concludes the posthuman pattern. It differs from any of the later transitions that comprise the Singularitarian pattern. The enhancement that occurs in the human pattern is good regardless of its relationship with the interests and desires of a child. It takes us from beings incapable of life plans to beings capable of them. Later enhancements do not have this effect. Whether or not they are good depends on how they relate to the life plans of the beings who are subject to them. One of the conclusions of this book is that typical life plans of humans entail a very restricted endorsement of enhancement.

5 Should We Enhance Our Cognitive Powers to Better Understand the Universe and Our Place in It?

This chapter addresses the value we place on improving our capacity to do science by enhancing our cognitive powers. The search for new scientific explanations is certainly not the only motivation for enhancing our cognitive powers—one might pursue cognitive enhancement for its effects on the writing of poetry or the playing of chess, for example. But it has a special significance. Scientific advances enable the invention of new technologies that improve our lives. In addition, they satisfy a deep and enduring curiosity about the universe and our place in it. This curiosity is held by some to be the defining virtue of our species.

In chapter 2, I used MacIntyre's distinction between internal and external goods to address the consequences of enhancement. This distinction serves as a useful starting point for investigating the effects of radical cognitive enhancement on science. Science produces many external goods. It reveals the causes of diseases enabling successful treatment of them. It enables the construction of jetliners that quickly and safely ferry us between continents. These satisfy MacIntyre's criteria for external goods. If there were ways of getting these goods that did not require science—perhaps by praying to a god who would just banish diseases from existence and delegate angels to facilitate air travel—then their value would be undiminished. The internal goods of science correspond with our sense of curiosity about the universe and our place in it. The scientific advances that enabled successful treatment of diseases and the construction of jetliners brought new understanding of the universe, understanding that divine miracles do not in general bring. This internal good of science is not restricted to practicing scientists. The publishing genre of popular science satisfies the curiosity of nonscientists, granting them access to this particular internal good. Prayers that bring the valuable external goods of cures for disease and angel-powered flight would not satisfy this sense of curiosity about the universe and our place in it. They promise none of science's internal goods.

In this chapter, I argue that the external and internal goods of science respond differently to cognitive enhancement. The propensity of cognitive enhancement to generate external goods conforms to the objective ideal. The greater the degree of objective enhancement, the more significant will be the external goods. By quickening the pace of scientific discovery, radical cognitive enhancement will accelerate the discovery of technologies that improve our lives. The more we cognitively enhance, the sooner we will receive beneficial technologies. Our cognitive faculties become more instrumentally valuable with greater degrees of enhancement.

The propensity of cognitive enhancement to generate science's internal goods conforms to the anthropocentric ideal. Beyond a certain point, greater degrees of objective enhancement tend to bring less valuable internal goods. They satisfy our sense of curiosity less effectively. Another way to put this is to say that our cognitive faculties tend to lose intrinsic value with greater degrees of enhancement.

The explanations of radically cognitively enhanced scientists (henceforth "radically enhanced science") differ from the explanations of unenhanced human scientists (henceforth "human science") by connecting less directly with the internal goods of human science. They respond less directly to the human sense of curiosity about the universe and our place in it. This difference arises in respect of the idealizations that constitute these sciences. Idealization is an indispensable feature of science—human scientists idealize, and so will radically cognitively enhanced scientists. But their idealizations will differ. These differences arise as a direct consequence of the respective cognitive limitations of unenhanced and radically enhanced scientists. One reason scientists idealize is to simplify reality so that they can formulate theories about it. Scientists with different cognitive powers will have different requirements for simplicity and should therefore favor explanations formulated in terms of different idealizations. Unenhanced human science is adapted to unenhanced human cognitive capacities. It lacks idealizations that exceed human cognitive limitations. This differs from a radically enhanced science featuring idealizations adapted to radically enhanced scientists' more capacious minds.

Unenhanced scientists are right to find greater intrinsic value in explanations that make use of idealizations adapted to their cognitive limitations than they do in explanations presented in terms of idealizations adapted to radically enhanced intellects. These explanations connect with a particular scientific narrative (or a particular collection of scientific narratives) about the universe and our place in it. They add to or revise a story about how the universe is and how it came to be this way that includes contributions by Aristotle, Copernicus, Newton, Darwin, Curie, Einstein, and other great scientists. I will argue that when scientists accept that their theories should appeal to idealizations constrained by the limits of human intellects they are not thereby required to accept a limit on what can be explained. They are not required to accept that there must be parts of the universe forever beyond their ken. There are, in fact, good grounds to believe that idealizations tailored to the limits of human minds need leave no aspect of the universe and our place in it beyond explanation by humans.

The preceding paragraph has addressed science's internal goods. Suppose that radically enhanced science connects less well with the internal goods of human science. The enhanced capacities would be less intrinsically valuable. This is so even if we recognize the great instrumental value of radically enhanced science. Radically enhanced science could give us spaceships capable of traveling at many times the speed of light and cures for all present and future diseases without satisfying our desire for explanations of the universe and our place in it. I will argue that radically enhanced science presents to us as revealed wisdom. We believe revealed wisdom about the universe because we recognize it as coming from beings or sources that have some privileged access to the truth. It is a feature of some revealed truths that they combine high instrumental value with low intrinsic value—they may enable us to do many things without needing to satisfy our sense of curiosity about the universe and our place in it.

If we underwent radical cognitive enhancement, then radically enhanced science would no longer present to us as revealed wisdom. Our much more intelligent future selves would find its proofs and demonstrations genuinely illuminating. They would genuinely satisfy *their* need to understand the universe and *their* place in it. This is what we should expect if we have correctly identified radical enhancement as a transformative change.

Must we choose between an intrinsically valuable human science whose instrumental value is comparatively low and a radically enhanced science that lacks intrinsic value but which possesses great instrumental value? Perhaps not. I conclude this chapter with a proposal that the enhancement of our cognitive powers is not the most effective way to gain science's external goods. The most effective means of pursuing these goods should focus on enhancing the tools that scientists use rather than on enhancing the scientists themselves. We can preserve science's propensity to satisfy our sense of curiosity while doing our best to translate its results into cures for diseases and vehicles that will carry us to the stars.

Understanding the Consequences of Cognitive Enhancement for Science

An interest in scientific progress provides strong *prima facie* support for cognitive enhancement. We do science to explain events and phenomena that take place in the universe. Certain types of cognitive enhancement should make us better scientists by enabling better explanations of these events and phenomena.[1] Most dramatically, cognitive enhancement might enable the discovery of explanations that could not be found by unenhanced human scientists. Less dramatically, it could accelerate the discovery of new explanations. This second interpretation of enhancement's benefits does not postulate explanations that are necessarily undiscoverable by humans. But one of its implications is that, supposing that we have a finite time to find out things about the universe, we will discover explanations that might otherwise have gone undiscovered. We should expect there to be some humanly knowable but forever unknown scientific explanations.

To understand claims about the degree of cognitive enhancement it's useful to postulate a specifically scientific ideal toward which cognitive enhancement could approximate. Those with an interest in scientific explanation would recognize different degrees of cognitive enhancement as standing at different distances from this ideal. This ideal is Laplace's demon, an intelligence imagined by the early nineteenth-century French astronomer and mathematician Pierre-Simon Laplace. Laplace describes the demon as

an intellect which at a certain moment would know all forces that set nature in motion, and all positions of all items of which nature is composed, if this intellect were also vast enough to submit these data to analysis, it would embrace in a single formula the movements of the greatest bodies of the universe and those of the tiniest atom; for such an intellect nothing would be uncertain and the future just like the past would be present before its eyes.[2]

Laplace's demon seems, on the face of it, to possess a capacity to explain and predict that could not be improved upon. There are at least three reasons that it may seem perverse to posit it as an ideal toward which enhancement of our capacity to do science should approximate.

It's clear that the demon knows many things of which human beings are ignorant. It knows all that can be known about the locations and behavior of nature's fundamental particles. But there are things of which we are quite knowledgeable about which it may be ignorant. We have some counterfactual knowledge. We know that a car parked on a slope with its handbrake properly applied will remain in place. But we also know what *would* happen were the handbrake not to be engaged. If the future presents to the demon exactly as does the past, then the demon may lack this understanding of counterfactuals. The demon knows that the car will remain in place, but doesn't know that it *would* roll down the slope were the brake not to be applied. Counterfactual knowledge is a central element of scientific explanation. Deprived of this knowledge, the demon knows that the car remains in place but cannot explain why it does. I propose that we remedy this possible epistemic deficit by crediting the demon with counterfactual knowledge. If the demon has a single formula that embraces "the movements of the greatest bodies of the universe and those of the tiniest atom," then a single presumably much longer formula describes all the actual and potential movements of the greatest bodies and tiniest atoms.[3]

Second, the demon may seem deprived of knowledge of higher-level phenomena about which humans are pretty well informed. For example, the demon makes predictions about a human being based on knowledge of all of the fundamental particles that constitute the human. It does not appeal to the higher-level entities—beliefs, desires, and other intentional states—that guide our predictions and explanations of behavior. By one measure, its explanations of human behavior in terms of microphysics are superior to our explanations in terms of folk psychology. Its complete knowledge of microphysics enables more accurate and complete predictions of future locations of human bodily parts than does our knowledge of folk psychology. However, suppose that folk psychological states such as beliefs and desires are genuine states of human beings. Should we suppose that the demon lacks knowledge about these states? If the answer to this question is yes, then we know some things that the demon does not know. One way to get around this difficulty is by adding perfect knowledge of higher-level states to the demon's stipulated perfect knowledge of microphysics.

These additions may seem to bloat the demon's mind to the point of absurdity. Remember that Laplace's story is a philosophical thought experiment, not some practical prescription for programmers of artificial intelligences. It serves as an ideal—a mere theoretical possibility—against which to measure a scientist's powers of explanation and prediction. Radically cognitively enhanced scientists should approximate more closely to Laplace's demon than do unenhanced human scientists.

Third, it may seem odd to present the demon as some sort of perfect scientist. The demon has no need for science. The starting point of Laplace's thought experiment is knowledge of all the "forces that set nature in motion, and all positions of all items of which nature is composed." To these data it applies "a single formula" which gives it perfect knowledge of the future. Beings who proceed from a partial picture of the universe use science to improve their capacity to explain and predict. It's a measure that they take to fill in gaps in their knowledge. It's possible that quantum uncertainties prevent the demon from perfectly predicting. But we should nevertheless recognize the demon as making predictions and finding explanations that are as good as any being could make. Science becomes dispensable only when a being achieves the level of knowledge about the universe attained by the demon. Consider a being whose epistemic powers are slightly inferior to those of the demon. This "almost-demon" lacks knowledge of an extremely small percentage of the universe's atoms. It could use science to eliminate this epistemic deficit. We can understand cognitive enhancement as getting us closer to the demon, even as we acknowledge that it will never enable the demon's powers of explanation and prediction.

We can locate radically cognitively enhanced scientists at some point between human science and the knowledge of Laplace's demon. They, like us, but unlike the demon, begin with a partial picture of the universe. Their more capacious minds lead them to approach the demon's epistemically perfect state at a faster rate than do our own more limited intellects.

There are differences among philosophers as to how science extrapolates from the partial picture of the universe presented by our senses. These differences actually matter little for the following discussion—this chapter's conception of science and scientific progress is sufficiently broad to encompass a wide range of philosophies of science. However, for expository purposes it will be useful to keep in mind a simple model of how scientists are supposed to come by new explanations. According to the influential philosopher of science Karl Popper, scientists encounter a phenomenon that they wish to explain.[4] They hypothesize an explanation. This hypothesis is then tested. According to Popper, when scientists test a hypothesis they seek to make observations or perform experiments that might falsify it. Popper's account of the scientific process does not arrive at certainty. The explanations offered by hypotheses that survive falsification are retained but remain open to future falsification and consequent expulsion from the body of explanations that comprises science.

Popper's account of the scientific method permits a clear statement of the case for cognitive enhancement. Enhancements of scientists' powers of imagination should allow them to come up with superior hypotheses. Other cognitive enhancements will improve scientists' abilities to formulate tests and to recognize when a given test falsifies a hypothesis. It seems difficult to motivate an anthropocentric condition for such value. Hence, the rationale for enhancing our cognitive powers seems to conform to the objective ideal. The more we cognitively enhance, the better we are likely to be at science and the closer we will be able to approximate to the knowledge possessed by Laplace's demon.

In what follows I explore an alternative, anthropocentric ideal for scientific advancement. There is a distinctively human science, conditioned in part by our distinctively human cognitive limitations. We place greater value on additions to and alterations of the body of explanations that constitute human science than we do on acquiring the scientific explanations of radically enhanced scientists. This is not to say that explanations of radically enhanced science have no value for humans. They do. They can have great instrumental value. It is to say that we have a justified preference for explanations that contribute to a scientific narrative (or scientific narratives) that comprises the explanations offered by Aristotle, Newton, Darwin, Curie, and Einstein. Radically enhanced science is properly viewed as different in kind from human science. Young scientists dream of contributing new explanations to that narrative, or they fantasize about successfully challenging extant explanations. Contributions to and modifications of that narrative genuinely advance human scientific understanding. They satisfy our sense of curiosity about the universe and our place in it. Human scientists are justified in valuing new explanations in accordance with the contribution that they make to their scientific understanding. The explanations of radically enhanced science will present to us as revealed wisdom. They tend not to satisfy the curiosity of human scientific explainers.

It might seem that acceptance of a science conditioned by human cognitive limits is effectively acceptance of explanatory mediocrity. By resting content with familiar varieties of explanation, we accept that the really deep truths about the universe and our place in it will remain forever beyond our ken. This is no necessary consequence. The fact that there might exist a radically enhanced science that is beyond human cognitive powers does not imply that there exist phenomena necessarily beyond the reach of human science. The possible existence of a radically enhanced science does not mean that human physicists cannot achieve a Theory of Everything, a theory that explains and links together all physical phenomena.

Two Ways in Which Human Science and Radically Enhanced Science Might Be Fundamentally Different

There are two ways in which human science could be fundamentally different from radically enhanced science. The most obvious way posits a difference *in truth*. Human scientific explanations differ from radically enhanced scientific explanations because the former point to human truths whereas the latter point toward distinct radically enhanced scientific truths. An explanation that radically enhanced scientific explainers would be right to view as true could be false for human scientists, and vice versa. Perhaps some truths will be shared. The rudiments of the theory of evolution might be true for both human and radically enhanced scientists. But this shared region is unlikely to contain all of the truths that radical cognitive enhancement would permit us to learn about the forces that construct and maintain life.

This view resembles a cultural relativism about scientific truth that awards to different cultures different scientific truths. According to cultural relativists about scientific truth, the fact that a statement appears on the list of scientific truths of one culture is no reason for it to be acknowledged as true by another culture.[5] Defenders of a species-relativism about scientific truth might be pessimistic about the prospect of receiving new scientific knowledge from advanced extraterrestrial civilizations. Humans and extraterrestrials might happen to have some scientific truths in common, but the fact that extraterrestrials rightly accept a given claim as true does not guarantee its truth for human scientists. The point at which extraterrestrial truths become too complex for humans to understand could be, on this approach, an appropriate place to separate extraterrestrial truths that are also truths for us from extraterrestrial truths that are not. Species-relativism about scientific truth might lead to pessimism about the benefits of radically enhancing our powers of scientific reasoning. It seems to lead to the conclusion that we have no business pursuing truths that exceed our current cognitive limitations.

Cultural relativism about scientific truth faces significant challenges. The existence of a shared world confronted by different cultures leads to skepticism about cultural relativism; so too, the existence of a shared universe inhabited by scientists, unenhanced, radically enhanced, and extraterrestrial should make us suspicious of a species-relativism about scientific truth. It seems odd to say that the act of radically enhancing a scientist's intelligence could falsify formerly true claims about subatomic particles. This chapter's discussion assumes a realism that assigns to science the purpose of uncovering truths about a mind-independent, objective reality. A scientific realism of this type leaves no place for scientific relativity about truth conditioned by the intellectual limitations of scientists. The difference in kind between human science and radically enhanced science cannot be a difference in truth. Obviously, there is much more that one could say on the topic of relativism about scientific truth. A paragraph's worth of observations constitutes no decisive refutation. Nevertheless, I now pass to another possible fundamental difference between scientists whose cognitive powers differ greatly.

Differences in Idealization as Fundamental Differences between Human and Radically Enhanced Science

Suppose that the truth or falsity of a claim about the universe is not contingent on a scientist's degree of cognitive enhancement. There's a further fact about scientific explanation that introduces the possibility of relativity. This fact depends directly on the cognitive capacities of the members of a scientific community. The explanations that comprise our science are likely to differ systematically from the explanations that constitute the science done by radically cognitively enhanced scientists.

One way to clarify the difference that radical cognitive enhancement should make to science points to the indispensable role of idealization in science.[6] There is heated debate among philosophers of science about the reason scientists idealize. One reason is to simplify a physical reality that is—in objective terms—hideously complex. According to a very helpful overview of this purpose of idealization offered by the philosophers of science Roman Frigg and Stephan Hartmann, "An idealization is a deliberate simplification of something complicated with the objective of making it more tractable."[7]

Later in this chapter, I discuss an additional reason for scientists to idealize. Advocates of this view present certain kinds of idealization as enhancing the power of scientific explanations. They achieve this enhanced power by omitting or distorting causally irrelevant detail. I propose that this interest in idealization is compatible with viewing idealization as also functioning to simplify reality.

Human scientific explainers have a powerful need to make reality simpler. Consider the job confronting scientists tasked with explaining human beings. According to one estimate, a typical human body comprises some 7×10^{27} atoms.[8] The properties of these atoms vary significantly. They enter into hugely diverse combinations with other atoms. The atoms are themselves composed of many different subatomic particles, each with their own distinct properties. Complexity is no obstacle to Laplace's demon who starts off both knowing everything that can be known about the intrinsic properties and possible interactions of the particles. But this complexity presents differently to human scientists who must simplify if they are to attempt to explain other human beings. We must idealize if we are to succeed in explaining each other.

Frigg and Hartmann identify two ways in which idealization can make reality more tractable. What they call Aristotelian idealization removes from a scientist's imagination the properties deemed irrelevant to his or her explanatory purposes. This stripping away of properties enables explainers to focus on "a limited set of properties in isolation." They give as an example of Aristotelian idealization the model in classical mechanics "of the planetary system, describing the planets as objects only having shape and mass, disregarding all other properties." Aristotelian idealizations omit aspects of reality to make it more tractable. Galilean idealizations deliberately distort reality to facilitate the task of explanation. Consider Boyle's gas law. This law distorts the molecules of a gas. It falsely asserts that they are perfectly elastic and spherical, possess equal masses and volumes, have negligible size, and exert force on one another only during collisions. The scientists who appeal to Boyle's law are not stupid. They are aware of the law's omissions and distortions. Scientists gain the benefit of its explanations and predictions without having to believe that there are such things as perfect gases—gases that correspond precisely and fully with Boyle's law.

Consider another example of deliberate simplification for the purposes of scientific explanation. When astrophysicists explain the motion of planets they know that there are many objects in the universe sufficiently large to exercise gravitational influence over celestial bodies. Suppose you're attempting to explain the motion of the Earth. A very simple idealization that enables a pretty accurate explanation considers only the Sun and the Earth. A more representationally accurate idealization includes the Moon. Still more accurate idealizations include all of the planets in the solar system. A model that includes only these objects omits the gravitational influences of other large objects in the universe. It ignores the gravitational effects of dark matter. It ignores the small but real influence on planetary orbits of quantum effects. A model that treats planets as perfect spheres omits the effects of topological contours. Our model of planetary motion involves Aristotelian and Galilean idealizations—it both omits from reality and distorts it. The result is a good (enough) explanation of the Earth's motion.

Idealization occurs at all levels in science. Our current models of quantum phenomena involve the Aristotelian maneuver of deliberately omitting information about subatomic entities and, in Galilean fashion, ascribing properties to them that we know they do not possess. Current models of economic behavior distort by ascribing perfect knowledge of markets to human beings. This false ascription enables predictions of large collections of human beings in response to economic incentives.

Earlier in the chapter, I proposed Laplace's demon as an ideal toward which cognitive enhancers who aspire to improve science might aim. Laplace's demon has no need for Aristotelian or Galilean idealizations. Its intellect is sufficiently large to encompass every single particle in the universe. Its explanations of gases and planets are, in an objective sense, superior to our own—the demon predicts aspects of these systems' behavior of which human scientists are ignorant. The limitations of our human intellects necessitate idealizations. The cognitive powers and knowledge of radically cognitively enhanced scientists will not carry them to the heights of predictive and explanatory power achieved by Laplace's demon. So they will require idealizations too. But these idealizations should differ from those used by human scientific explainers. A scientific community whose members have intellects significantly more powerful than the intellects of the most intelligent human scientist is likely to have less need to sacrifice predictive and explanatory accuracy to make reality cognitively tractable.

A community of scientists negotiates its own compromise between the benefits of idealizations that are representationally accurate and the requirement that scientific ideas be sufficiently simple for its members to grasp and freely manipulate in thought. A community of more cognitively gifted scientists will find a different point of compromise between idealization and accurate explanation from that found by a community of less cognitively gifted scientists.

We can place the idealizations of different scientific communities on a scale that measures their representational accuracy. Laplace's demon uses idealizations that are perfectly representationally accurate, that is, they are not really idealizations at all. The demon's representations of reality include all features that make any actual or potential difference to any process that it seeks to understand. The being that Laplace imagines has no need to economize in its representations of reality. We should expect the representations of cognitively enhanced scientists to be more representationally accurate than those of unenhanced human scientists.

Consider how differently able communities of scientists approach the challenge of explaining human behavior. Laplace's demon takes into account all of the approximately 7×10^{27} atoms and massively more numerous subatomic particles that make up a human being. This attention to detail enables it to explain things about human beings that scientists forced to rely on less representationally accurate idealizations cannot. Our explanations respect human cognitive limits. Belief-desire psychology seems to be a paradigm case of a theory that achieves explanatory breadth by simplifying its subjects. We would never dream of attempting to explain an entire human brain in terms of the interactions of the subatomic particles that comprise it. Consider radically cognitively enhanced scientists undertaking to explain human behavior. They fall short of the explanatory completeness of Laplace's demon. There is likely to be a different point of compromise between their requirements to simplify and their interest in explanations that are as complete as possible. Explaining an entire human (or posthuman) brain in terms of its fundamental physical constituents might be beyond them. But perhaps they might attempt explanation in terms of the approximately 100 billion neurons that comprise human brains or the around 8,000 genes expressed in the brain.

Scientists who self-consciously idealize appreciate that their simple models do not exactly describe every detail of reality. They should not view the existence of more complex theories as falsifying their own more simple conjectures. According to Kepler's first law of planetary motion, planets trace perfectly elliptical orbits about the Sun. Whether or not this law is a good one depends, in part, on the uses one wishes to make of it. It's fine for amateur astronomers seeking guidance on where to point their telescopes. It would, however, be inadequate for NASA engineers seeking to plot the course of a rocket they hope to place into orbit around Uranus. Discrepancies that amateur astronomers are right to view as negligible assume great significance for NASA engineers. We can only guess at the interests in explanation of radically enhanced science. But it's reasonable to suppose that they will find significance in some discrepancies between idealization and reality that we justifiably view as negligible.

Idealizations That Enhance the Power of Scientific Explanations

My discussion so far has focused on the role idealizations play in simplifying reality to make it tractable by human explainers. Radically cognitively enhanced scientists will lack the knowledge of Laplace's demon. They too must simplify reality. But their need for simplification is less extreme than our own. Hence the explanations that comprise radically enhanced science will differ from those that comprise the science of unenhanced humans.

Simplification is not the only need that philosophers of science imagine idealization fulfilling.[9] According to Michael Strevens, idealizations enhance the power of scientific explanations by permitting the omission of details about reality that make no causal difference to phenomena that we are seeking to understand.[10] He objects to the notion that "our weak intellects cannot bear reality in all of its radiant, complex profusion, so we water down our explanatory models with simplifying falsehoods to achieve something inferior but more cognitively palatable."[11] Strevens presents an alternative view according to which idealization improves the quality of explanations. In this view, idealization makes explanation more powerful by isolating the "causal difference-makers" from total collections of causal influences. He takes an example from Wesley Salmon to demonstrate that some causal influences are not causal difference-makers:

A ball goes astray during a local baseball game; it hits a window and the window breaks. Among the causal influences on the window are both the ball and shouts of the players, the latter because sounds cause the window to vibrate. But of these two influences, only the ball makes a difference to the breaking. Why? Both affect what happens to the window as it breaks. The influence of the shouting is subtle; it perhaps affects the precise pattern of the shattering, hence the exact shapes of the shards and their exact trajectories. But these influences do not (except under very unusual circumstances) steer the trajectory of the window so as to determine whether it breaks or not. They make a difference to the way that the window breaks, but not to the fact that it does break. Thus they do not explain the breaking.[12]

In this case, idealization improves explanation by permitting the omission or distortion aspects of reality—the shouting—that are causal influences, but not causal difference-makers. Earlier I offered Boyle's gas law as an example of idealization's simplification of reality. Strevens provides an alternative account of the law's omissions and distortions. He suggests that by omitting or distorting factors, forces, or tendencies that are not difference-makers, the law focuses attention on causal difference-makers. "When the explanation of Boyle's law, for example, falsely assumes that gas molecules do not collide, it is telling you that the collisions in a real gas make no difference to the fact that it conforms (approximately) to Boyle's law."[13]

The idea that idealization can do the job Strevens describes does not prevent it from also performing the role of simplifying reality. This simplification is evident not so much in how we understand particular explanations, but rather in the kinds of explanation that members of different scientific communities tend to seek. Human and radically enhanced scientists should agree that the ball caused the window to shatter. But the radically enhanced might have a legitimate interest in "the precise pattern of the shattering, hence the exact shapes of the shards and their exact trajectories." Human observers are unlikely to view these aspects of the window's breaking as properly eligible for explanation. Our idealizations present reality to us in a form fit for integration into explanations that we can understand. But there are presumably properties of the ball, the window, and the baseball stroke that are causal difference-makers for the window's shattering into exactly 487 shards, each one composed of a specific number of atoms. These could feature in the explanations of radically enhanced science.

Boyle's law is good enough for unenhanced human explainers of gases. It focuses attention on causal difference-makers that lie within human comprehension. The truth of this account is compatible with the existence of difference-makers at the atomic and subatomic levels. Presumably the precise atomic and subatomic properties of gas molecules affect their collisions with other gas molecules. Beings with an interest in these collisions will presumably need to appeal to idealizations with greater representational accuracy than Boyle's law.

Mathematics as a Bridge between Human and Radically Enhanced Science

Have I really shown that the idealizations of radically enhanced science are beyond the cognitive limits of unenhanced human scientists? An awareness of the history of science reveals that our idealizations are not static. There is a tendency for more representationally accurate idealizations to replace less representationally accurate idealizations. This may seem to suggest that there is no discontinuity between the idealizations of unenhanced scientists and those of radically enhanced scientists.

Consider the history of improvements of our understanding of the orbits of planets. According to Copernicus, the planets trace perfectly circular orbits around the Sun. This idealization had great value. It explained in a simple way much of what had been mysterious about the observed motion of the planets. It was, however, soon superseded by the less distorting idealization of Kepler, according to which planets trace elliptical orbits. From a scientific realist perspective, this is clearly an improvement. Kepler's model of a planet's orbit corresponds more closely to reality than does that of Copernicus. There are limits on the degree of improvement that it is possible for human scientists to achieve. Laplace's demon explains the orbits of planetary bodies in ways that do not idealize at all. Its predictions proceed from a complete account of every particle in the universe. We can presume that the replacement of more distorting explanations for less distorting ones will stop short of the demon's explanatory powers.

Or is this so? Perhaps humans have access to a cognitive tool that could, in principle, help them to achieve the demon's level of scientific competence. A significant part of the story about the replacement of more by less distorting explanations points to mathematics. Mathematics enables the description of the somewhat distorting Copernican circular orbits. It enables the description of Kepler's less distorting elliptical orbits. More complex mathematical models take into account influences overlooked both by Copernicus and Kepler. The "single formula" that Laplace's demon uses to explain "the movements of the greatest bodies of the universe and those of the tiniest atom" makes use of mathematics. Does this, in principle at least, place it within reach of human scientists equipped with mathematical understanding?

The utility of mathematics as a bridge between human science and the superior science of more cognitively able beings features in a recent book, *The Eerie Silence*, by the physicist Paul Davies. Davies explores the possibility of exchanges of information between humans and super-intelligent extraterrestrials. He is enthusiastic about what we might learn through communicating with extraterrestrial super-intelligences. He says, "Being in touch with ET would expose our civilization to accumulated cosmic wisdom, and open the way to technological marvels, deep scientific insights and entry to the Galactic Club."[14] In particular, Davies would like help from ET with "some famous problems like how to bring gravitation and quantum physics together, the long-sought-after theory of quantum gravity."[15] More on that problem later.

Davies views mathematics as solving another problem that threatens our acquisition of knowledge from ET. How might humans open communications with extraterrestrials? There would be no shared language that they could use to convey this valuable information. How could a conversation with ET even get started? Davies recommends that we use mathematical concepts and truths to open communications. According to Davies, it's the universality of mathematics that would make it a good choice:

Mathematics occupies an unusual place in our culture in that it is a product of the human mind, and yet it transcends the mind. Any sufficiently advanced being elsewhere in the universe could prove the same theorems based on the same logical principles. Given that the universal laws of physics are manifested in the form of elegant mathematical regularities, it is clear that mathematics is the key to bridging the gulf between human and alien cultures. If aliens know any science, or have developed any advanced technology at all, then they will be familiar with mathematics. They will even be familiar with the same mathematics as we know.[16]

I suspect that mathematics might help to solve the problem of how to open communications with super-intelligent extraterrestrials. For example, we might understand their query "Do you inhabit the *third* rock from the Sun?" as containing the number concept 3. But a tool that serves as an excellent conversation starter is likely to disappoint as a means for them to convey their more complex scientific theories.

Mathematics may fail to serve as a bridge between human science and the sciences of cognitively superior beings because there are likely to be mathematical truths that scientific explainers can use in constructing their models of the universe that are beyond human understanding. There are infinitely many mathematical truths. The cognitive limits of human scientists will prevent them from understanding or being able to effectively reason with some of these truths. These mathematical truths will therefore be unavailable for the construction of models of the universe. More intelligent extraterrestrial scientists are likely to have access to a larger subset of these mathematic truths than do their human counterparts. They are likely to draw on these in the construction of their scientific models. At least, there's no reason that they should draw on the same parts of mathematics that human scientific explainers do. Their bigger minds will give them a greater tolerance for complexity—they won't require models that omit from or distort reality to the same degree that our models do. This greater tolerance should apply at all levels of extraterrestrial scientific explanation.

Human Science, Radically Enhanced Science, and the Theory of Everything

Suppose that it is the case that radically enhanced science is beyond comprehension by unenhanced humans because it makes use of idealizations at least some of whose complexity is beyond us. In this section, I ask what might be gained through undergoing radical enhancement, leading to a gradual exchange of scientific explanations using idealizations adapted to human cognitive abilities for explanations whose idealizations are adapted to radically enhanced minds. My particular focus will be on the internal goods of science—science's capacity to satisfy our curiosity about the universe and our place in it. The discussion's principal example is an example of scientific knowledge that we do not yet possess but that we can predict would lead to very valuable internal goods. This knowledge would do a great deal to satisfy our curiosity about the universe and our place in it.

The Theory of Everything looms large on the agendas of theoretical physicists who seek to take the long view of their discipline. Some scientists and commentators on science view it as a kind of ultimate goal for theoretical physics—a theory that in some significant sense explains what the universe is all about. According to the physicist Steven Weinberg, the Theory of Everything will explain and link together all physical phenomena.[17] It would enable the in principle prediction of the result of any experiment. This ability to predict *in principle* differs from Laplace's demon's capacity for universal perfect prediction.

One of the current obstacles for those who aspire to a Theory of Everything arises in respect of quantum gravity. Physicists are now working hard to unify quantum mechanics, which comprises three out of four of the "fundamental interactions"—electromagnetism, strong nuclear force, and weak nuclear force—with gravity, as described by the theory of general relativity. This is proving enormously difficult.[18] The assumptions about the universe made by the theory of general relativity seem irreconcilably different from those of quantum mechanics. Theoretical physicists are currently exploring a few promising leads. Some hope to find the answer in string theory.[19] Others are skeptical.[20]

The Theory of Everything would enable very valuable internal goods of science. We might justly view it as a tragedy for our species if we turned out to be clever enough to understand that there could be such a thing as Theory of Everything, yet just too dumb to ever know it. So, would our collective need for a Theory of Everything justify some degree of cognitive enhancement? In what follows, I argue that radical cognitive enhancement may have a self-defeating effect on the quest for a Theory of Everything. As our cognitive abilities become more powerful, an adequate Theory of Everything should become more difficult to discover. Hopes for a Theory of Everything may depend on forgoing cognitive enhancement.

Dawkins and Haldane versus Deutsch on the Limits of Human Science

Before we ask whether cognitive enhancement is likely to help us find a Theory of Everything, we should ask whether we are likely to find it without undergoing cognitive enhancement. It's possible to identify two views about possible limits on the scientific knowledge of unenhanced human scientists. We can contrast an optimistic view of their powers due to the theoretical physicist David Deutsch with a pessimistic view due to the biologists J. B. S. Haldane and Richard Dawkins. The former view acknowledges no need on science's behalf for cognitive enhancement. Indeed, Deutsch goes so far as to assert that cognitive enhancement could make no difference to our ability to discover fundamental scientific truths. The latter view concedes that there could be phenomena that succumb to the explanatory powers of enhanced scientists but not to those of the unenhanced.

In his recent book *The Beginning of Infinity*, Deutsch challenges the idea that there is a strict limit on what humans can find out.[21] The "infinity" of his book's title is meant as a rebuke to those who propose that we, as finite creatures, must resign ourselves to a finitude of knowledge. Deutsch rejects a view about human cognitive limits that he calls the Haldane–Dawkins argument—Haldane for the English biologist J. B. S. Haldane who originally proposed it, and Dawkins for Richard Dawkins, a recent popularizer of the idea. The Haldane–Dawkins argument is summarized by Haldane's famous dictum "the universe is not only queerer than we suppose, but queerer than we can suppose." Dawkins gives this claim an evolutionary twist.[22] Our cognitive faculties evolved to enable our survival and reproduction. According to Dawkins, this largely confines us to a "middle world" made up of entities larger than quarks and atoms but smaller than galaxies. Knowledge of these middle-world entities made systematic differences to our ancestors' abilities to find food and mates, avoid getting eaten, and successfully raise children. Our cognitive faculties depart the middle world only with great difficulty. To suppose that they could reveal facts about the universe irrelevant to evolutionary success is perhaps like supposing that our fingers, perfectly adequate for the purposes for which they were designed—moving around the kinds of middle-world objects that make a difference to survival and reproduction—should be up to the task of shifting individual atoms around. The farther we travel from the middle world, the more likely we are to find things and processes incomprehensible by humans.

Deutsch strongly rejects the idea "that progress in science cannot exceed a certain limit defined by the biology of the human brain. And we must expect to reach that limit sooner rather than later. Beyond it, the world stops making sense (or seems to)."[23] He opposes the pessimism of Haldane and Dawkins with a view of human beings as "universal explainers." Deutsch presents the Enlightenment as the event in human history that turned humans into universal explainers. We exchanged explanations with necessarily limited reach for ones whose reach is not limited. We made a "jump to universality." Parochial explanations—explanations whose finite reach means that they apply only to a narrow range of familiar circumstances—gave way to explanations with infinite reach. The explanations of early twenty-first-century physics have infinite reach. Our current theories about hydrogen cover not only hydrogen atoms that we have experienced directly, but any hydrogen atom at any time, at any place. They are true of an infinity of actual and potential hydrogen atoms. Deutsch thinks that the unlimited reach of human explanations casts doubt on the very idea of significantly enhancing our powers as explainers. He thinks that human minds are already universal: we "can create knowledge about anything." "Since humans are already universal explainers and constructors, they can already transcend their parochial origins, so there can be no such thing as a superhuman mind as such."[24]

Deutsch's view supports some confidence on behalf of unenhanced human scientists in search of the Theory of Everything. If Haldane and Dawkins are right, then the Theory of Everything could be beyond our cognitive reach. The central elements of such a theory are likely to be located well outside of the middle world for which our cognitive and perceptual faculties evolved. In Deutsch's view, our capacity for universal explanation could be an evolutionary by-product. But now that we have it, we can apply it to all manner of phenomena both relevant and irrelevant to human survival and reproduction.

I suspect that whether we are actually smart enough to ever discover a Theory of Everything remains an entirely open question. There's no reason to suppose that an adequate Theory of Everything must be within the reach of human minds. We could be universal explainers yet unable to discover a theory that adequately explains and links together all physical phenomena. It follows that if we are universal explainers then some of our scientific explanations will have infinite reach. The scope of these explanations will not be restricted to objects and processes that are present to our senses. Rather, they will apply to all object and process of certain types. There's nothing in this that requires that these will be *good* explanations with infinite reach. For example, the theory that each living being is surrounded and deeply affected by an aura, a field of luminous, spiritual energy, can be interpreted as having infinite reach. Its advocates should expect to find aura fields around all living beings, terrestrial and extraterrestrial. Aura theory is deficient as science, but not because of the parochial nature of its claims. Consider a particularly dull subpopulation of humans. The members of this subpopulation are universal explainers in Deutsch's sense.

They understand perfectly well that if uncontaminated water on Earth can be safely consumed, then uncontaminated water anywhere and anytime should be safe to drink. Their explanations take the same form of those of civilizations capable of discovering the Theory of Everything. But there's a good chance that they, working individually or collectively, wouldn't ever come up with an adequate Theory of Everything.

On the other side, Dawkins errs in supposing that the mere fact that our minds are products of evolution shows that aspects of the universe alien to our immediate experience must or are likely to be beyond our cognitive reach. Some conceivable discoveries about the design history of our cognitive faculties might make us confident that certain types of knowledge were beyond us. For example, suppose that we were to discover that our cognitive faculties had been designed by an all-powerful yet shy creator who was determined to prevent us from acquiring knowledge about it. We might conclude that no use of our cognitive faculties would help us to understand our creator's nature. Recognizing that natural selection had no interest in our being able to formulate a true Theory of Everything supports no such skepticism. It's improbable that members of our species ever gained an evolutionary advantage through specifically *not* being able to arrive at good scientific theories about objects and processes that are irrelevant to our biological fitness. Human scientists have discovered good accounts of many aspects of the universe that are entirely irrelevant to our evolutionary flourishing. Knowing that californium-251 has a half-life of around 2.64 years could conceivably make a difference to human reproductive fitness—if, for example, you found yourself held hostage by a Trivial Pursuit–obsessed psychopath. But it would be odd to describe its discovery as boosting scientists' reproductive fitness. There's no reason to suppose that the members of a scientific community able to discover obscure facts about transuranic elements could not also come up with a good Theory of Everything.

I propose that neither considerations about the evolutionary origins of human cognitive faculties nor those about the human status as universal explainers provide sufficient information to say with any confidence whether some arbitrary as yet undiscovered piece of knowledge, such as the Theory of Everything, could be acquired by humans. We know that the Theory of Everything will be beyond beings who are not universal explainers. We don't know whether we are universal explainers who are capable of discovering it.

How Different Idealizations Generate Different Theories of Everything

I have presented unenhanced human science as differing from radically enhanced science by virtue of appealing to less representationally accurate idealizations. As we have seen, some progress in science occurs with the discovery of idealizations that more closely mirror reality. There is nevertheless likely to be an upper limit on the representational accuracy of idealizations tolerable by human minds. Beyond this upper limit, there are likely to be theories of the universe whose idealizations are more representationally accurate than are those accessible to human minds. I claim that this does not require that there may be significant parts or aspects of the universe that are forever beyond the reach of human explainers.

We can make sense of cognitively enhanced scientists being, in some significant sense, better explainers than us, while at the same time insisting that there could be no limit on the phenomena that unenhanced scientists can explain. The more capacious minds of radically enhanced scientists can contain idealizations that are more representationally accurate than are those that our unenhanced minds can contain. Unenhanced scientists are confined to less representationally accurate idealizations. But the mere fact that these idealizations are less representationally accurate does not imply that they cannot yield good scientific explanations of any phenomenon in the universe. Good unenhanced human science genuinely explains phenomena and enables accurate predictions. To translate this claim into Deutsch's terms, a comparative lack of representational accuracy does not mean that unenhanced scientists cannot be universal explainers, wielding explanations with infinite reach. There need be no physical phenomenon in the universe that we cannot discover a good theory about.

Perhaps we will find good explanations of currently perplexing events inside black holes. These good explanations will answer all the questions that unenhanced human scientists will ever ask about black holes. But these good explanations will involve idealizations conditioned by the limits of human intellects. The fact that more intelligent beings would prefer explanations that make use of more representationally accurate idealizations does not, in itself, invalidate our explanations. The fact that they find our theories inadequate does not compel us to echo them.

Consider an analogy with maps. Mapmakers once used the labels "terra incognita" and "mare incognita" to indicate regions about which little or nothing was known. Modern maps of the world dispense with this label. They are complete in the sense that they have global reach—there are no regions marked as beyond the knowledge of the mapmakers. We might consider global reach to be a junior sibling of Deutsch's universal reach. But to attribute a global reach is not to say that there is no information about the world omitted from them. The fact that our best world maps omit information does not require us to view them as incomplete. For our explanatory purposes they are complete. A ship's captain heading into a region marked on her map as "mare incognita" should not hope for any navigational assistance from the map. A modern map offers as much information to ships' captains about the most distant bodies of water as it does about the seas that abut their home ports. If this map is the product of unenhanced human cartographers, it will make use of human carto- graphical conventions, conventions conditioned by human cognitive and perceptual limitations. It will omit information about the world that car- tographers predict could be of no value to those who use the map. We should not view the map as incomplete just because it does not use carto- graphical conventions adapted to radically enhanced cognitive and per- ceptual abilities.

It's possible that humans are insufficiently clever to arrive at a Theory of Everything. We will remain forever clueless about how the fundamen- tal forces of electromagnetism, strong nuclear force, and weak nuclear force might relate to the fundamental force of gravity. But suppose that humans are sufficiently intelligent to formulate and find scientific proof for a theory that explains and links together all physical phenomena. This human Theory of Everything should differ from a Theory of Everything that makes use of the more representationally accurate idealizations of radi- cally enhanced science. It is unfair to deny unenhanced human science a claim on the label "Theory of Everything" just because there could exist a theory that uses more representationally accurate idealizations to explain and link together all physical phenomena. A comparatively representation- ally inaccurate human Theory of Everything could do exactly what human scientists expect from it. It could successfully explain and link together all physical phenomena *for us*.

Denying to unenhanced humans a true Theory of Everything risks deny- ing the possibility of a true Theory of Everything to radically enhanced scientists. After all, the idealizations of radically enhanced science lack the perfect representational fidelity of those of Laplace's demon. Remember that the demon's perfectly representationally accurate Theory of Everything is constituted by a single formula that precisely describes the possible and actual behaviors and interactions of every single fundamental particle in the universe. We can grant radically enhanced science a Theory of Everything even as we allow that the idealizations that compose it are less representationally accurate than are those of Laplace's demon.

Might unenhanced human scientists suffer the tragedy of being intelligent enough to know that there could be a Theory of Everything, while being just too stupid to ever hypothesize and confirm one? There is a threshold of cognitive sophistication that members of a scientific community must exceed to be able to have any Theory of Everything. We can't know *a priori* if unenhanced human scientists exceed this threshold. But there are some grounds for confidence. A survey of debate among theoretical physicists seems to reveal plenty of new hypotheses. String theory has brought many new explanatory strategies to the quest. We certainly aren't at the stage yet at which physicists are throwing their hands up in the air and passing the problem over to philosophers.

The radical cognitive enhancement of human scientists may do nothing to bring them closer to a Theory of Everything. Rather, it is likely to significantly change what would count as a Theory of Everything. The more representationally accurate idealizations of radically enhanced science are likely to make something that they would accept as an adequate Theory of Everything significantly more complex than a Theory of Everything that uses idealizations adapted to human cognitive limitations. We may find that our very attempts to dramatically increase the explanatory power of our science increase the difficulty of achieving our explanatory aspirations. Radical cognitive enhancement is likely to be self-defeating if conceived as a means of reducing the overall quantity of scientific mystery.

Science is finished for Laplace's demon. It explains everything by means of perfectly representationally accurate idealizations. But there are likely to remain scientific perplexities for beings who do not achieve the explanatory perfection of the demon. I think that this is a good thing. Neither we nor our cognitively enhanced descendants are likely to lose the distinctive variety of questing that is characteristic of science. The unanswered scientific questions of future civilizations will differ from ours, but there are unlikely to be fewer of them. As the idealizations of our enhanced descendants become more representationally accurate the questions we use them to ask become more difficult to answer.

This may make science seem a Sisyphean task in which no matter how hard we try we never make any progress. But this would be a false impression. Scientists will continue to ask and answer fundamental questions about the nature of the universe and our place in it. Scientific progress will continue to generate internal goods—it will enhance understanding of the universe and the questioners' place in it. In addition, it will generate external goods. As the representational accuracy of idealizations increases, we will be able to build more powerful technologies. The technologies of radically enhanced beings may propel them beyond our galaxy at speeds faster than light. It's good to know that they will still be driven by a sense of curiosity about the universe and their place in it.

Valuing Human Science and Radically Enhanced Science

So far I have said little about the value we attach to science. I have argued for a collection of factual claims. The science practiced by unenhanced humans is likely to differ in kind, not merely in degree, from that practiced by radically cognitively enhanced scientists. Left to our own unenhanced devices, human scientists are unlikely to produce the same explanations as those that comprise radically enhanced science. This need not mean that human science is necessarily incomplete. The mere fact that it appeals to idealizations that are within the limits of our intellects does not prevent us from arriving at a Theory of Everything. It need not leave any phenomenon in the universe necessarily beyond our explanatory reach.

Suppose we have the power to radically enhance our cognitive powers. This would permit us to do radically enhanced science. Should we do this?

We can ask this question in two ways. First, we can ask about the effects of this enhancement on the intrinsic value of our cognitive faculties. How would this change affect these capacities' propensity to yield science's internal goods? How well would it satisfy our need to understand the universe and our place in it? Second, we can ask about the effects of this enhancement on the instrumental value of our cognitive faculties. How would this change affect these capacities' propensity to yield science's external goods? To what extent could it improve human lives by curing disease and enabling interstellar travel?

Radical cognitive enhancement is a transformative change. It reduces our connection with the internal goods of human science. It makes science less able to satisfy a distinctively human interest in understanding the universe and our place in it. We can allow that radical cognitive enhancement would, if successful, improve the instrumental value of our cognitive faculties. It would increase the expected yield of external goods. But this conditional fact alone does not establish that radical cognitive enhancement is the best way to pursue science's external goods. We most efficiently pursue science's external goods by enhancing the tools that human scientists use rather than the human scientists themselves.

Radical Enhancement Reduces the Intrinsic Value of Our Cognitive Faculties

What attitude should we take to the science that radical enhancement may enable us to perform? Consider how we should think about the same science if it were produced not by our radically enhanced future selves but by super-intelligent extraterrestrials.

Suppose that super-intelligent extraterrestrials do overcome Davies's translation problem and communicate some hitherto unknown scientific claims to us. Imagine that we're convinced of the sincerity of the extraterrestrials. We observe their stupendous technologies and infer that they are much more intelligent than us. We have good reason to believe the scientific truths presented to us by extraterrestrials. But this belief may not be accompanied by genuine understanding. The extraterrestrial explanations may make use of idealizations whose complexity is beyond human cognitive limits. We may be incapable of grasping the mathematics that the extraterrestrials use to describe their idealizations—the mathematics may come from regions beyond human cognitive limits. Humans may justifiably believe the conclusions of extraterrestrials' scientific arguments without understanding why they might be true.

The science fiction author Arthur C. Clarke famously said, "Any sufficiently advanced technology is indistinguishable from magic." A working high-definition television would present to a group of Pleistocene hunter-gatherers as magic. They could have no grasp of the various interacting technologies that produce the images. I propose a version of Clarke's claim specific to science. Any sufficiently advanced scientific explanation is indistinguishable from revealed wisdom. Unenhanced humans should believe that the sincerely presented statements of radically enhanced science are true. But our reasons for believing these claims differ from our reasons for believing claims presented by unenhanced human scientists. We believe them because we recognize them as sincere presentations of beings in an epistemic position far superior to our own. We understand that the idealizations that the radically cognitively enhanced scientists offer as evidence are beyond the limits of our understanding—they aren't and couldn't be evidence for us unless we were to undergo radical cognitive enhancement. For us, their explanations are not and could not be science. They neither require genuine human understanding nor advance it.

There are two ways to discover knowledge about the universe and our place in it. One is science. Earlier I described a broadly Popperian approach to science according to which scientists formulate and seek to falsify testable hypotheses. Hypotheses that survive this process are retained. Another way in which we might discover truths is by revelation. Revealed wisdom comes from a source or agent that has superior access to relevant knowledge. The term "revealed wisdom" is most familiar from religious contexts. According to some, the being with the best possible access to truths about the universe and our place in it is an omniscient and omnibenevolent god. God or gods are not the only sources of revealed wisdom. What's essential is the superior epistemic position of those who provide us with the information—our justification derives from beliefs about the provenance of the information. We believe it because we believe it to come from a being or beings who should know better than us. Revealed wisdom might come from super-intelligent extraterrestrials. Unenhanced humans may come by it from cognitively enhanced humans. Parents reveal wisdom to their children. Children should believe parental warnings about the dangers of busy roads without having to seek further proof.

We often have good reason to believe information presented as revealed wisdom. When we do, our justification does not point to a fact or collection of facts about the universe in virtue of which the claim might be true; rather, it points to the epistemically superior status of the being or beings who conveyed the information.

Some ways of gaining new knowledge about the universe and our place in it bring no new understanding. Consider the answer to the "Ultimate Question of Life, the Universe, and Everything" presented in Douglas

Adams's novel *Hitchhiker's Guide to the Galaxy*. Adams imagines a civilization whose desire to answer the Ultimate Question leads them to construct a supercomputer—Deep Thought—to find the answer. After seven and a half million years, Deep Thought delivers the somewhat unsatisfactory answer "42."

There's been much discussion about why this answer is disappointing. I'll explore one explanation. The problem with 42 as the answer is not that it's false. It should, in Adams's fiction, be accepted as true—given that it comes from a cognitive agent far smarter than those to whom it's presented. We can suppose that Deep Thought is well designed and appropriately programmed. Adams's petitioners should accept that it is correct. "42" is an unsatisfactory answer when presented to humans because it's not appropriately connected with the body of truths that constitute human understanding of the universe and the meaning of life. Even when we're assured that the answer is 42, knowing this fact does not enhance our understanding. We can't possibly see how it could be true. Perhaps a radically enhanced intellect could see, after being presented with the reasoning, that the answer couldn't be 41 or 43 but just has to be 42. But for us, "42" is no better or worse an answer than "41," "43," or "pterodactyl!" None of these potential answers connect with the concepts that we use to frame our inquiries about life, meaning, and the universe. Adams acknowledges this fact when he imagines the recipients of the answer seeking to build an even more powerful computer (which happens to be the Earth) to determine the Ultimate Question. One presumes that the Ultimate Question would connect 42 with our current philosophical and scientific inquiries into life, meaning, and the universe.

The hope for scientific explanations enabled by radical enhancement is, from our present perspective, the same as wishing for revealed wisdom. We are hoping for explanations that do not connect with the explanations that constitute our current scientific explanations. This revealed wisdom is less intrinsically valuable than are explanations that genuinely extend human understanding. This is part of the intrinsic value we attach to advancing scientific understanding. Advancing scientific understanding is more than just adding new truths. It continues a series of narratives whose key figures include Aristotle, Copernicus, Newton, Darwin, Curie, and Einstein. Young biologists might enjoy imagining presenting their discoveries about natural selection's mode of operation to Darwin. In this thought experiment, they present a summary of their reasoning to the father of evolutionary theory. This thought experiment makes no sense if they imagine themselves practicing radically enhanced science. The discoveries enabled by radical enhancement do not belong to human scientific narratives. As such, they do not genuinely advance human understanding of the universe and our place in it. They belong to entirely different scientific narratives.

The human scientific narrative is not a simple thing. Einstein challenged much of Newton's conception of the universe—he didn't just add further elements to the Newtonian narrative. But Einstein's challenge took a form intelligible to human physicists. He offered alternative idealizations of the workings of the universe that key members of the physics community could understand. He did not seek to change scientists' minds about the nature of the universe by merely announcing his conclusions and offering the results of cognitive aptitude tests confirming his superior epistemic status. When human scientists present their results they desire to present them in ways that are intelligible to human scientists. If what I have said above is true, we have no reason to think that the possible existence of a scientific narrative that draws on more complex idealizations leaves the human scientific narrative necessarily incomplete. Suppose that Deutsch's claims about the future of human science are correct. Human science may still have infinite reach. There may be no event or phenomenon in the universe that resists explanation in its terms. To return to the map analogy, the maps of human scientists need have no areas marked "terra incognita." They may one day be complete. But they will be complete in their own terms. The fact that the maps of radically enhanced cartographers pick out key features in a way that cleaves more closely to their objective structure does not give the lie to these claims.

What of Scientific Enhancement's Instrumental Benefits?

The claim that we humans place greater intrinsic value on advances to our powers of scientific reasoning does nothing to show that radically enhanced science is without value. At the beginning of this chapter, I distinguished the intrinsic value of a new scientific explanation—its propensity to generate science's internal goods, to satisfy our sense of curiosity about the universe and our place in it—from the instrumental value of a new explanation—its propensity to generate science's external goods, to improve our lives by leading to new technologies. We place greater value on explanations that genuinely advance human understanding than on explanations that we may have good reason to be true, but which do not advance human understanding.

The enhancement of our cognitive faculties could increase their instrumental value. It would accelerate our discovery of beneficial technologies. In chapter 3, I argued that the radical enhancement of human cognitive capacities is likely to be an indirect, inefficient way to pursue external goods. The same points apply to our cognitive capacities' generation of useful knowledge. If we are serious about finding cures for cancer and building safe nuclear fusion plants, then we should seek to enhance the technologies that scientists use to generate new discoveries in these areas rather than seek to enhance the scientists themselves.

This chapter's discussion introduces a different reason to forgo the radical enhancement of our cognitive faculties as a means of tackling difficult practical problems. I have suggested that radically enhanced science differs in kind from human science. Both sciences describe the same universe. But radically enhanced science is likely to proceed along a line parallel to human science, drawing on idealizations beyond the cognitive limits of the practitioners of human science. Suppose we were to decide to expedite our search for a cure for cancer by radically enhancing the intellects of cancer researchers. This option essentially scraps the human inquiry into cancer, restarting with a radically enhanced approach informed by idealizations beyond the understanding of the human sciences currently grappling with the disease. This might be a good idea if we had reason to believe that our current avenues of research were unlikely to lead to a cure for cancer, or at least to something sufficiently close to a cure so that the disease loses its power to traumatize individuals and families. In effect, we would cease trying to solve the problem of cancer, redirecting our effort toward the quite different problem of how to radically enhance the cognitive powers of scientists with the expectation that they would then rededicate themselves to the problem.

This seems an unduly pessimistic evaluation of our progress against the disease. Cancer scientists lack a cure, but their attempts to find one have taught us a great deal about how the disease begins, how it spreads, and how it can be combatted. This is the picture that emerges from the account of cancer presented in Siddhartha Mukherjee's "biography" of cancer, *The Emperor of All Maladies*. Mukherjee tells a story of many disappointments, due largely to a systematic underestimation of the complexity of cancer, leading to a much richer account of the multiple environmental and genetic causes, conditions, and triggers of the disease. It's difficult to say how far we might be from something we might call a "cure" for cancer. But it seems we have made nonnegligible progress toward one. To the extent that we are truly interested in finding a cure for cancer, we should direct our attention at the problem at hand—the problem of how to cure cancer.

Powerful technologies currently extend the cognitive reach of cancer researchers. Enormous quantities of data on people who do or don't get specific types of cancer are being input into computers that discern patterns obscure to human observers. The minds of human cancer researchers tend to be limited to a few of the most obvious linkages. Computers extend the creativity and lateral thinking of human researchers by applying simple pattern-matching algorithms to vast data sets. These computer data-mining techniques enable the identification and exploration of linkages that are unlikely ever to be spotted by a human cancer researcher. Consider one problem about to benefit from the enhancement in creativity and imagination enabled by data mining. We know that cancer results from a hideously complex interaction of almost uncountable environmental factors including diet, exposure to microbes, chemicals, sunlight, and so on, with a vast array of hereditary factors. In the past, we've had to rely on the intuition of researchers to intuit a possible connection between a cancer and an environmental or hereditary factor. Such a "hunch" led to the studies that confirmed smoking as a significant cause of cancer. Data miners are now able to explore a—to human minds—bewildering range of connections between cancer and environmental and hereditary factors. Perhaps they will uncover a small increase in the rate of melanoma for people with a certain genotype who weren't breast fed, ate small quantities of brassicas but relatively large quantities of tuna sashimi, and spent their early years living in cold climates. These advances do not require any improvements to human brains. They emerge instead from the entirely uncreative number-crunching of the increasingly powerful computers used by cancer scientists. This approach has the additional benefit of preserving intrinsic value of our own thought processes while pursuing external goods of great instrumental value.

In this chapter, I've used the idea of an idealization to distinguish human from radically enhanced science. Scientists whose intellects are radically more powerful than our own are likely to understand the universe in ways that differ from the ways in which we understand it. There is likely to be a difference in kind, not merely a difference in degree, between human science and radically enhanced science. We place a reduced value on explanations produced by this radically cognitively enhanced science. We rightly view it as not addressing our curiosity about the universe and our place in it. This confession of cognitive inferiority does not entail a renunciation of our ambition to completely understand the universe and our place in it.

6 The Moral Case against Radical Life Extension

The longest verified human life span is that of the smoking, drinking Frenchwoman Jeanne Calment, who lived for 122 years and 164 days. Calment died in 1997. She had vivid memories of meeting Vincent Van Gogh—"a dirty, badly dressed, disagreeable" man. Calment put her extreme longevity down to a diet rich in olive oil, port wine, and chocolates.[1] Radical life extension would give its recipients life spans that far exceed that of Madame Calment. Aspiring radical life extenders want to live for thousands of years.

There's a difference between the topic of this chapter, radical life extension and the topics of the previous three chapters, the radical enhancement of our physical and intellectual powers. The explicit purpose of the latter is to change us in quite significant ways. Physically enhanced runners will record marathon times out of the reach of unenhanced human athletes. Radically cognitively enhanced scientists will formulate theories far beyond the intellects of unenhanced human scientists. Radical life extension seems to have the reverse goal. It promises to protect us from change—specifically from age-related change. If all goes according to life extenders' plans, you in a thousand years' time may differ little from you now. We might therefore expect few of the barriers of imaginative identification that thwart our engagement with radically cognitively or physically enhanced beings.

It is possible to object to the prudential rationality of radical life extension. In my earlier book, *Humanity's End*, I mounted such an objection.[2] Enthusiasts present multiplying a conventional human life span by ten as multiplying by ten all of the good experiences in a conventional human life. I argued that this is unlikely to be so. The activities that seem to us to be rewarding and pleasurable—travel to exotic locations, driving sports cars, viewing the Niagara Falls from close up, and so on—are likely to seem too dangerous to our especially long-lived future selves. We should view radical life extension as a transformative change that purges our lives of our familiar pleasures. This chapter offers a different kind of response to life extension. I argue that translating the theoretical possibility of radical life extension into a therapeutic reality is likely to require immoral acts. This charge of immorality does not attach directly to the act of extending one's life. For the purposes of this chapter's argument, we can allow that the act of acquiring a millennial life expectancy is, in itself, morally permissible. But the experiments necessary to make life extension a therapeutic reality will impose significant unjustified and undeserved burdens on others. I derive this conclusion from two different claims. The first claim concerns the science of radical life extension—it addresses the experiments necessary to establish that a proposed therapy safely halts or reverses aging. The second kind of claim concerns the motivations of those who desire radical life extension. I argue that very few individuals who seek radical life extension will freely offer themselves as participants in clinical trials for anti-aging therapies. Since they are unlikely to volunteer in the numbers required, the search will be on for others to serve as anti-aging guinea pigs.

I present the enterprise of radically extending human life spans as requiring an immoral transfer. The central characters in vampire stories achieve radically extended existences by draining the blood of victims. In this act, "life force," or something similar, is transferred from the human victim to the vampire. The concept of life force has no place in modern biology, and this means that radical human life extenders simply could not arrange its transfer from disempowered donors to empowered recipients. The empowered do nevertheless acquire something from the disempowered that has consequences similar to supping on their blood—this is their participation in dangerous clinical trials. The empowered will achieve their extended life spans by effectively taking years from others. The mechanism by which these years are transferred differs from that in vampire stories, but its consequences are the same.

Two Kinds of Anti-Aging Research

There are two kinds of anti-aging research. The traditional approach focuses on *diseases of aging*. In this chapter, I shall understand diseases of aging as comprising the diseases acknowledged in conventional medical practice as both becoming more prevalent as humans age and predictably leading us to die sooner than we otherwise would have. I will refer to diseases of aging as DoAs. Included on the long list of DoAs are Alzheimer's disease, atherosclerosis, type 2 diabetes, cancer, and a host other life-shortening ailments that become more common as we grow older. The threat that DoAs pose to human flourishing is widely appreciated. The fight against them consumes vast sums of money. Those researching DoAs seek both to extend human life expectancies and to enhance the quality of the last stages in these lives.

Then there's the more ambitious approach, which sets as its target *aging as a disease*, or AaD. Its goal is an end to aging. The ultimate goal for those targeting AaD is not just a few additional, comparatively healthy years. This research aims at an end to human aging. Ageless humans should have average life spans of around a thousand years. They could live much longer if they managed to stay out of the way of misdirected buses and errant meteorites.

Many popular presentations minimize what must be done to treat AaD effectively. They return to the myth of the fountain of youth in which a single intervention—a few sips of water—restores youth. For example, in Drew Magary's 2011 book *Postmortal: A Novel*, a scientist discovers, quite by accident, "the cure for aging." Three simple injections decisively end the aging process. A single cure for aging may serve the purposes of fiction, but it is highly unlikely to come to pass. Realistically, an end to aging will require multiple therapies each targeting a distinct variety of age-related damage.

The dissident gerontologist Aubrey de Grey offers the most complete current plan to treat AaD. De Grey's Strategies for Engineered Negligible Senescence (SENS) program aims to engineer negligible biological aging.[3] Negligibly senescent people will not age. It is a general fact about aging beings that each year that we live diminishes the number of years we should rationally expect to survive. Hale and hearty forty-year-olds should expect, on average, to die sooner than hale and hearty twenty-year-olds. With each year of life, the probability of dying in the next year increases, meaning that death soon becomes a statistical inevitability, with Jeanne Calment left as a truly remarkable statistical outlier. Ageless humans will be exempt from this generalization. Negligibly senescing forty-year-olds should expect the same, very large number of remaining years as negligibly senescing twenty-year-olds.

De Grey does not minimize what must be done to achieve negligible senescence. It will require quite profound revisions and repairs of the features of human biology that allow the ongoing accumulation of age-related damage. De Grey's SENS program is searching for therapies that will intervene in the aging process in more extensive and profound ways than any drug that shrinks Alzheimer's plaques or helps people with diabetes to better regulate their blood sugar, the kinds of therapies sought by researchers on DoAs. De Grey's road map to agelessness requires that we address seven fundamental causes of aging—the "seven deadly things." I documented the key elements of de Grey's proposal in my book *Humanity's End*, so I will be sparing in my presentation of it here.

The focus of the SENS strategy to end aging differs from the more superficial approaches that attract media attention in the early twenty-first century. In de Grey's view, strategies such as taking a daily jog, frequently hugging a pet, consuming dietary supplements, or rigidly adhering to Madame Calment's dietary regime of olive oil, port, and chocolate can have only modest effects because they do not target the "fulcrum of aging."[4] These superficial approaches can only be stop-gaps as we await the development of approaches that address the fundamental causes of aging.

De Grey thinks that an interest in the fundamentals of human aging should direct our attention to the level of the cell. He proposes that the vast catalog of diseases of aging recognized by contemporary medicine reduce to seven problems with the internal workings of cells or the ways in which they relate to each other. As we age, we tend to lose cells in places that we need them. To give two examples, the processes of cell replacement do not compensate for the loss of cells in crucial organs such as our brains and hearts. As this is occurring, other parts of our bodies acquire cells that are both superfluous and obstruct proper biological functioning. One reason for the impaired response of older people to infection is that their immune systems are burdened with the maintenance of large populations of barely functional immune cells that would, ideally, be terminated, clearing the way for fresh fighters of infection. Mutations of DNA located inside the cellular nucleus and outside in mitochondria—the cells' power-plants—are further deadly things. Cancer is a particularly tragic consequence of the tendency of our cells to acquire genetic mutations. Mutations to certain key stretches of DNA end up killing by permitting a cell to divide without limit. The bodies of older people deal less efficiently with waste. It accumulates both within cells and between them. Alzheimer's disease is a consequence of the buildup of junk in sufferers' brains. Extracellular crosslinks are a special kind of junk that accumulates between cells. Hypertension is a harmful consequence of the tendency of crosslinks to make our artery walls less flexible. These cellular and intercellular mishaps are "the seven deadly things" that are the focus of de Grey's revolutionary brand of anti-aging medicine.

The SENS Response to the Seven Deadly Things

SENS undertakes to fix each of the seven deadly things. De Grey's 2007 book, *Ending Aging: The Rejuvenation Breakthroughs That Could Reverse Human Aging in Our Lifetime*, cowritten with Michael Rae, contains quite detailed suggestions about how each could be repaired or prevented.

For an example of one of the more fantastical elements of SENS, consider de Grey's proposed cure for cancer. Cancer is a deal-breaker for would-be life extenders. The incidence of cancer significantly increases with age, meaning that those who make it to 150 before there is a cure or at least a very effective treatment would be advised to cultivate a taste for chemotherapy. De Grey is a proponent of WILT, or whole-body interdiction of the lengthening of telomeres, which are the protective sequences of DNA at the ends of chromosomes. The first step in WILT involves excising from every cell in the human body the gene that produces telomerase, an enzyme that extends telomeres. The shortening of telomeres that occurs with each cell division seems to impose a strict limit on the number of divisions a cell-line can go through. Many cancers acquire the capacity to grow without limit by hijacking the telomerase gene—lengthening telomeres so that they no longer act as checks on cell division. Removal of the gene would require that incipient cancers reinvent from scratch the gene or some other means of indefinite growth. This is something that cancers are unlikely to be able to do—switching an existing gene back on is a much easier task than building it *de novo*. But de Grey's cure for cancer comes with a significant cost. Some parts of the human body churn through cells at a pretty rapid rate and therefore require an active telomerase gene. Without a well-functioning telomerase gene, we'd soon run out of red blood cells, white blood cells, and platelets. De Grey proposes that stem cell therapies supply the missing cells. In his vision of an ageless future, patients would front up for regular infusions of new blood cells and cells for our stomach linings. These new cells would combine long telomeres with the absence of a telomerase gene.

WILT is an eccentric approach to cancer—it differs quite markedly from traditional approaches. This, in itself, is not a bad thing. Cancer is an immensely challenging medical problem best confronted by a research community that divides its attention among an array of potential treatments in the knowledge that many of them will lead nowhere. At this point in humanity's engagement with cancer, we are well advised to follow Mao Zedong's advice and let "a hundred flowers blossom and a hundred schools of thought contend." The history of medicine includes many examples of breakthroughs in understanding that originated as eccentric proposals. However, I think it is a mistake to view WILT as just another oddball approach to cancer that may succeed where traditional approaches have failed. It has goals that differ markedly from the approaches to cancer typical of conventional medical research. Conventional medical research has the goal of either keeping people free of cancer over the duration of a normal human life span or ensuring that any cancers that they do contract do not kill them before they reach the end of this span. A cancer diagnosis is always bad news. But it may not be so bad if the diagnosis is followed immediately by the information that your cancer is growing so slowly that you will die with it, not of it. One line of research seeks to distinguish fast-growing cancers, those requiring the promptest action, from slow-growing ones that can be let be. A SENS cancer therapy aims to protect not merely over the duration of a normal human life span, but over a millennium. Even the slowest-growing tumors are bad news for an aspiring millenarian. There's further bad news from the fact that a significant component of cancer risk increases exponentially. Cancer becomes more common as we age. This is partially a consequence of copying errors to accumulate with each cell division. An eighty-year-old's cells have undergone more divisions than a twenty-year-old's. An eight-hundred-year-old's cells will have undergone vastly more cell divisions than an eighty-year-old's. All of this makes cancer a statistical inevitability, something waiting for us all should we live long enough.[5] An effective SENS cancer therapy must alter this fundamental feature of human biology. Its goals are thus considerably more ambitious than those of conventional medical research, which aims to keep people free of cancer long enough to complete a full human life and to prevent any cancers that we do contract from killing us before we achieve the three score years and ten plus some which we view as a human entitlement. These differences matter for the clinical trials that de Grey will need to arrange.

SENS is, to say the least, controversial.[6] It's important to recognize what is at issue in the debate over the possibility of conclusively treating AaD and thereby radically extending human life spans. It's clear that de Grey cannot end aging today. But the true claim that we cannot *yet* repair or prevent any of the seven deadly things does not lead to the claim that it is impossible for us *ever* to find effective therapies for AaD. If an end to aging is achievable—and there's no reason to think human agelessness violates some fundamental law of physics—then the sooner we start looking for it in earnest, the sooner we're likely to find it. The greatest innovation that de Grey has brought to the life extension debate is not some new proposed therapy. Rather, it's his redefinition of success in the quest to cure aging. De Grey sets as our goal *longevity escape velocity* (LEV). Just being alive tends to damage human brains and bodies much in the way that regular driving wears out car components. Damage builds up until it reaches critical levels, that is, we die. We will achieve LEV when we can reliably fix the damage at least as fast as it accumulates. You finish each year with a longer life expectancy than you had at the year's outset. It's not necessary to wait for a "Cure for Aging Found!" headline to herald the arrival of millennial life spans. Look instead for an accelerating succession of stories announcing proven compounds that combat a wide range of causes of aging. We're not there yet. But de Grey and his coterie of wealthy investors hope that LEV will arrive within the next thirty years. The preliminary goal of LEV helps to place the apparent absurdity of WILT in its appropriate context. De Grey views WILT as among the most challenging elements of SENS. The first significant steps toward LEV are likely to occur in other areas.

To hurry this process, de Grey helped set up the Methuselah Foundation, which offers cash prizes to researchers who break records in mouse longevity. The record as of January 2012 is 1,819 days, which is really not bad for an animal with a life expectancy in the wild of under a year. Unusually long-lived mice could awaken people to the possibility of radically extending human life spans in much the same way that Dolly the sheep inspired fears and hopes for human cloning.

De Grey has written persuasively about the effects of the pro-aging trance—and the possible consequences of eliminating it. He describes the pro-aging trance as a refusal to take seriously the possibility that we might actually end aging. De Grey concedes that the pro-aging trance can be rational. Indeed, he allows it has always been rational right up until the time that he set out the central principles of SENS. If you know that something horrible is happening to you and you can't do anything about it, then it is best to (try to) accept it. You should train yourself to think as positively about it as the constraints of your psychology permit. This applies to people condemned to life imprisonment without the possibility of parole, and it applies also to every aging human being. De Grey envisages the creation of a Methuselah mouse making the pro-aging trance both untenable and unnecessary. He presents the demise of the pro-aging trance as having a powerful effect on the politicians of wealthy democracies. Aging will come to seem like terrorist attacks on passenger jets—something that is not only preventable, but worth trying very hard to prevent. The governments of wealthy democracies will commit substantial tax revenues to fighting AaD or face electoral defeat.

The following discussion uses de Grey's SENS program as its principal example. SENS is certainly the most completely elaborated current approach to the features of human biology that entail aging. My criticisms should apply to any approach that, like de Grey's, aspires to do much more than give us slightly more active eighth and ninth decades.

Is Aging Really a Disease?

Does de Grey's classification of aging as a disease have any basis in reality? If aging is a disease, then the popular expression "healthy aging" is as much an oxymoron as would be "healthy congestive heart failure."[7] Could this really be so? The pioneering bioethicist Daniel Callahan finds several differences between aging and disease. First, aging, unlike disease, is "a 'natural' biological inevitability." Callahan says, "unlike other pathologies, none of which is inevitable in every person, all humans are subject to it, as is every other organic creature"; furthermore, "aging is predictable, that is, in a way that nothing else ordinarily classified as a disease is. We may, or may not, get cancer or heart disease or diabetes, but we will surely get old and die."[8] Old age causes disability in ways that differ from the forms of disability brought on by disease. Callahan says, "much of the decline associated with age, particularly the increase in chronic disease and disability, is accessible to cure or relief," and he continues: "Even many of the other biological indices of aging—decline of hearing, rise of blood pressure, bone mineral loss, reduced muscle mass, failing eyesight, decreased lung function—are open to compensatory intervention though not at present to complete reversal."[9]

Callahan does successfully distinguish aging from *some* diseases. The combination of complaints that we group together under the label "old age" kills in ways that differ from that of metastatic cancer. But some diseases have the features that Callahan attributes to the purportedly nondisease process of old age. Some degree of dental decay exists in any human mouth that has ever come into contact with food. Many diseases cause gradual decline that, like healthy aging, is amenable to compensatory intervention. Type 1 diabetes is unmistakably a disease. Yet diabetics can influence the course of their disease by controlling their blood glucose levels, taking up exercise, and improving their diets.

We do think of aging as a process that brings disability somewhat gradually. But suppose we follow standard practice in bioethics and theoretical medicine and define a disease as a disruption of biological functioning.[10] This way of defining disease makes considerations of the speed with which a biological system breaks down irrelevant to deciding whether or not this breakdown is a disease. The malfunctioning prostates of men with prostate cancer are in a disease state regardless of how quickly or slowly the disease progresses. Some men diagnosed with prostate cancer learn that their tumor is growing so fast that it will soon kill them. Others learn that their tumor is growing so slowly that they would have to live until 150 for it to cause significant problems. The second group is certainly more fortunate than the first. But both groups of men have been diagnosed with a disease.

An interest in the philosophical accounts of disease seems not to distinguish the micro-injuries that comprise each category of de Grey's deadly things from the macro-injuries—heart disease, Alzheimer's and so on—that your local doctor would readily recognize as a disease. Each of the seven deadly things is a departure from normal biological functioning at the micro-level. Our cells were designed by natural selection to function individually and relate to each other in certain ways. The seven deadly things are failures to function and relate in these ways. Of course, taken individually, they don't cause any of the symptoms we associate with disease—a single malfunction in mitochondrial DNA produces no noticeable effect. The micro-injuries produce impairments in gross functioning that we more readily acknowledge as diseases only collectively. Considered individually, they are silent, symptom-free diseases like very slow-growing prostate cancer.

In what follows, I allow the micro-injuries associated with de Grey's seven deadly things to count as diseases. They are departures from normal biological functioning. They often end up having bad consequences for those who suffer them. This concession does not bar us from finding differences between diseases of aging and aging as a disease. These differences are not located at the level of biological fundamentals. Rather, they lie in our attitude toward them—more specifically, our willingness to accept the costs associated with the development of SENS therapies. I predict that people are more willing to accept the risks involved in clinical trials of therapies for conventional diseases than they will be to accept risks involved in clinical trials of SENS therapies. This is so even when both are diseases—they are disruptions of proper function that end up having bad consequences for those to whom they occur.

The Testing Problem

De Grey wants to make quite radical modifications to the ways our cells function and relate to one another. His focus on the fulcrum of aging differs from taking a new dietary supplement and hoping to notice an extra spring in our steps. At some point, new anti-aging therapies will have to be tested on subjects who are relevantly similar to those for whom they are intended. It's in the area of testing that we find differences between research on DoAs and research on AaD.

Consider the process of recruiting human subjects for a new therapy for a disease of aging. It's not too difficult to entice those suffering from DoAs into clinical trials. For example, people with Alzheimer's recognize that they have a terrible illness. They understand that many experimental drugs don't work—that some may actually make them sicker. Furthermore, they understand that a great therapy for Alzheimer's can make them sick in a variety of ways. Increases in LDL cholesterol and risk of stroke are warranted if the therapy does a good enough job in retarding the advance of Alzheimer's disease. Of course, if you're dying of Alzheimer's, you prefer to have the proven cure—but failing that, you're quite keen to be involved in a clinical trial of a therapy that might actually do some good. The main fear that people with Alzheimer's who present for clinical trials is that they'll end up in the trial's placebo arm and hence not receive the drug or therapy that trials on animals suggest may halt the advance of their disease.

People with serious DoAs often view the risks of enrolling in trials for untested therapies as warranted. They sometimes freely choose to risk their remaining months or years to have an improved chance at the life span and quality of life that we think of as a human birthright. The statistics make this consistent interest a bit puzzling. The widespread willingness of people with DoAs to participate is perplexing in light of the expected returns from such participation. A mere 8 percent of drugs submitted for approval to the US Food and Drug Administration are approved. There's some variation across different human diseases. The therapeutic hit rate for some diseases is higher than that for other diseases. But suppose that we take the 8 percent figure. This means that 92 percent of tested therapies fail because they are revealed to be ineffective or cause harms of greater magnitude than any therapeutic benefits. The 8 percent figure almost certainly overstates the expected benefits of an experimental therapy. Some therapies approved by the FDA are subsequently withdrawn because harms are discovered after they are approved. For example, Rofecoxib, a pain medication most widely marketed under the brand name Vioxx, was on the market for over five years before evidence of increased risk of heart attack and stroke led to its withdrawal. We should expect the statistics to be worse for therapies whose mode of action is genuinely novel. The therapies sought by de Grey fall into this category.

The low hit rate of clinical trials does not show that people are necessarily irrational to participate in them. But it does tell us something about their motivation for participating in such trials. It shows the high value that people place on normal human life spans.

Consider the implications of this widespread feeling of an entitlement to a normal human life span in the light of the frequently tragic, but in the end triumphant, history of pioneering work in open-heart surgery. Before the first successful procedures in the 1950s, many observers thought that surgery to repair defects inside hearts was a medical impossibility.[11] They reasoned that no patient could survive the stopping of the heart that would be required to repair it. Almost all of the earliest human subjects for pioneering work on open-heart surgery died as a direct consequence of the experimental procedures. What is especially tragic is that many of them were young children. In the early days, those aware of the obstacles in the way of successful open-heart surgery ought to have been aware that the expected return for a sick child was negative. Without the experimental surgery, sick children might expect to survive for some additional months with their gradually failing hearts. The universal failure of early attempts to surgically repair children's hearts should have made it rational to believe that an impending experimental procedure would also fail, killing the child. Perhaps this new procedure improved upon some of the established deficiencies of earlier procedures. But the great challenge of open-heart surgery ought to have made it unduly optimistic to think that this particular putative improvement would make the decisive difference. The experimental subject would be very likely to die, either on the operating table or soon after. The experimental procedure had a negative expected return—it was rationally expected to shorten a child's life. Set against this is the fact that the experimental surgery represented the only real opportunity for children with significant congenital heart defects to enjoy a normal human life span.

Were the parents of the children who participated in the earliest clinical trials of open-heart surgical procedures mistaken to give their consent? The trials do seem to have predictably shortened their children's lives.

According to one interpretation, the parents were making a mistake. On this interpretation, the parents were the victims of an established glitch in human rationality that social psychologists call loss aversion. Humans are more strongly motivated to avoid losses than they are to pursue gains. We will do more to avoid losing $100 than we will to gain the same amount. Banks take advantage of this defect in human rationality when they describe users of credit cards as missing out on cash discounts rather than incurring fees. People are more willing to use their cards when they view themselves as missing out on a possible gain than having to endure a penalty. The parents adopted a perspective according to which their children were threatened with a significant loss—the loss of the normal human life span that was their proper entitlement. Their attempts to make good this loss led them to subject their children to risky procedures that reduced the amount of life that it was rational for them to expect. They would be less interested in risky medical procedures that offered the same probability of giving their children years in excess of the normal human life span.

But linking a given choice to a psychological tendency to want to avoid losses does not suffice to show that the parents' choice was irrational. On a second interpretation, the parents were acting rationally. Suppose that the parents of the children in the open-heart surgery trials placed great value on their children's having a normal human life span or something close to it. They placed much greater value on any intervention that gave their children a chance at a normal human life span than on one that might extend that span beyond normal. The children's failing hearts doomed them to short lives. Participation in the trial represented the only realistic possibility of a normal human life span. This pattern of preferences could make participation rational in spite of an expected reduction of life span. I suspect that this kind of reasoning is behind the decisions of many very ill people who enter clinical trials fully aware that the experimental treatments are highly unlikely to benefit them and, in fact, likely to shorten their lives. They place great value on the possibility, small though it may be, that an experimental therapy will restore to them something like the normal human life span which they view as a human birthright. They recognize that any experimental intervention is quite likely to make them sicker.

I believe that the parents were justified in valuing a normal human life span as they did. The small chance that an experimental open-heart surgical procedure would actually succeed justified reducing their children's life expectancies—the amount of life it was rational to expect for their children. However, in what follows I do not rely on any normative claim about the importance of a normal human life span. This seems a suspiciously swift way to argue against SENS—we should distrust de Grey because he is offering us something other than the normal human life span. Rather, I limit myself to making a prediction about the choices that future people will make. Even if we suppose that placing such a high value on not falling short of a normal human life span is an instance of loss aversion, and is therefore irrational, it will significantly influence the deliberations of people whom de Grey wants to entice into his clinical trials.

The reference point of a normal human life span motivated parents of sick children to enroll them in clinical trials. I suspect that this reference point will have an opposite effect on the willingness of people to participate in the clinical trials that SENS requires. As with experimental open-heart surgery, the potential gains are great. The children and their parents were hoping for normal human life spans. De Grey offers the possibility of a millennial life span. Both open-heart surgery and WILT represent quite radical departures from existing medical practice. The first human experiments will be dangerous. In the case of open-heart surgery, people take risks in the hope of getting the normal human life span that we tend to view as our entitlement. In the case of WILT and other SENS therapies, candidates will be expected to risk their normal human life spans. If, as with open-heart surgery and many novel approaches to serious diseases, the expected return of initial SENS clinical trials will be negative, I predict that they will be especially averse to the possible loss of their normal human life spans.

Why WILT (and Other SENS Therapies) Will Require Dangerous Human Trials

At some point in the future, there may be safe SENS therapies. How could they be made safe? The purpose of the earlier section was to show that conventionally healthy people are likely to have expectations of a novel therapy that differ from those of people diagnosed with a serious illness. Members of the latter group tolerate negative expected returns in the hope that a therapy will grant them a normal human life span. Members of the former group are, in contrast, being asked to venture their normal human life spans. They are, as a consequence, likely to be significantly less tolerant of risk. They will expect a novel therapy to be shown to be safe before they submit to it.

How would we show that experimental anti-aging therapies are safe enough for people who satisfy all of the conventional requirements of health? If you are going to undergo a new therapy, you want it to have been extensively tested. You want these tests to have taken place in (preferably many) clinical trials on patients who are relevantly similar to you. If you have type 2 diabetes, you want medications that have been thoroughly tested on other people with this condition. Conventionally healthy people will want an anti-aging therapy to have been extensively tested on other conventionally healthy people. Conventionally healthy forty-year-olds will want to see the data on conventionally healthy forty-year-olds. I doubt that testing that does not involve this category of subject will suffice to encourage a conventionally healthy person to put at risk his or her anticipated normal human life span.

Testing on animals? The bulk of the experimental work on anti-aging therapies is done on nonhuman subjects. Mice are favored subjects. They, like us, are mammals. Conveniently, their life expectancies are short, which means that scientists don't have to wait decades to witness successful life extension. We should be alert to what experimental work on mice can, and cannot, show. Methuselah mice may be effective as proofs of concept—demonstrations that the life expectancies of mammal species including humans can be dramatically extended. But they certainly won't show that any *particular* therapy for AaD is safe. There are many biological differences between mice and humans; a therapy that carries a mouse to its 3,000th day could swiftly kill a human.

Testing on sick people? De Grey and Rae's book *Ending Aging* makes creative use of experimental work on DoAs and other recognized diseases. I suspect that this work should fall short of the standard of proof expected by the conventionally well. Consider a drug proven to extend the lives of people with Alzheimer's. What questions should an aspiring life-extender who has not received an Alzheimer's diagnosis ask before taking the drug? It could be that the drug extended the life of Alzheimer's patients by repairing micro-injuries also present in any aging brain. Or perhaps the drug is inefficacious when levels of this damage do not exceed a critical level, that is, a level sufficient for a diagnosis of Alzheimer's. There is an important difference between these two possibilities. A successful therapy for Alzheimer's may cause harms that do not prevent it from being a good therapy. Alzheimer's is a very serious disease. A drug that is good for people with Alzheimer's can have numerous ill effects that are collectively more than made up for by the therapeutic shrinking of the amyloid plaques. If the second explanation is true, then people whose level of amyloid plaques does not exceed a certain level will receive no benefit to offset these ill effects. There's an additional problem. People suffering from conventional DoAs are unlikely to be particularly interested in participating in SENS trials. They want immediate relief from the conditions that are preventing them from enjoying a normal human life span. They are, as a consequence, likely to be less interested in really ambitious SENS therapies that promise more but, realistically, have a lower likelihood of working. They are, in effect, being asked to sacrifice any opportunity to participate in another trial whose overriding purpose is to give them a normal human life span. For obvious reasons, patients enrolled in one clinical trial of an experimental therapy cannot concurrently enroll in a trial of another experimental therapy.

Testing on old people? Perhaps people reaching the end of their normal human life spans might be enticed into tests of experimental anti-aging therapies. They presumably have less to lose from therapies that do not live up to any initial promise. There are similar issues here to those that apply to testing on sick people. The bodies of older people have accumulated a large amount of age-related damage. Younger people might wonder whether a therapy that reverses some fraction of this damage would have the same effect on their less damaged bodies. Moreover, I suspect that old people might be reluctant to participate in SENS clinical trials. They are, in all likelihood, too far gone to be granted negligible senescence should the trial succeed. Why should they risk what they are entitled to view as a successful completion of their life's work? In an ancient Greek story retold by the playwright Aeschylus, Admetus, king of Pherae, is told he can be reprieved from death on the day he is fated to die, so long as he can persuade someone to freely take his place. It seems obvious to him that one of his elderly parents should make the sacrifice. They, however, demur, unwilling to give up the remainder of their allotted years. I think many older people will echo the sentiments of Admetus's parents. I doubt that many healthy seventy-year-olds will be tempted by a suggestion that since their time's almost up they should be prepared to play the role of a SENS guinea pig. Ninni Holmqvist's novel *The Unit* offers a particularly chilling vision of a future in which people who achieve a certain age without any socially valuable attachments or roles find themselves coerced into eventually lethal medical experiments.

Rational followers of SENS who are in early middle age, and conform with current norms of health, should demand data on people who are relevantly similar to them. De Grey needs to convince forty-year-olds who may be feeling pretty smug about having kept their cholesterol and blood pressure within medical norms that they should enroll in clinical trials of anti-aging therapies that reach into and modify the inner workings of their cells. I suspect that they are unlikely to want to sacrifice decades of good health to help rejuvenation researchers to eventually come up with a successor therapy that may actually work. What does this mean for WILT? It's possible that WILT is the cure for cancer. Suppose that it's not. WILT could still lead to hypotheses that really do cure the disease. It will be difficult to present this information in a way that would persuade people who satisfy the norms of conventional healthy aging to risk their remaining years.

Where to Find Human Guinea Pigs for SENS

If SENS therapies cannot be adequately tested, then that's the end of de Grey's plan to radically extend human life spans. The problem of where to find healthy, easily motivated human test subjects for SENS is actually depressingly easy to solve. We inhabit a world of vast inequalities. Some people have great wealth. Others live in extreme poverty. The poor will readily take risks and accept conditions rejected by the wealthy.

There are currently many cases of exploitation of the poor by the wealthy. Poor people are sex-trafficked. They work in sweat shops. They sell their internal organs. A widespread belief in the possibility of millennial life spans is likely to create an unprecedented demand for the services of the poor. I suspect that de Grey is right about the pro-aging trance and the consequences of its ending. If SENS does produce Methuselah mice or some other equally effective proof of concept, then people will come to recognize radical life extension as a real possibility. People will cease to view aging as something best addressed by poetry or existentialist philosophy. There will be pressure to turn SENS into a therapeutic reality as soon as possible. Any delay could make the difference for a middle-aged supporter of SENS between a merely acceptable eighty-year life span and a vastly preferable millennial life span. Currently Porsches and mansions are sold to the rich with the proviso that "you can't take them with you." They are likely to be enticed by the prospect of indefinite life spans that permit them to spend indefinite lengths of time with their Porsches and mansions.

One obstacle to realizing SENS is financial. If de Grey is right, then the demise of the pro-aging trance will see affluent democracies compelled by people-power to find the monies necessary to fund SENS. Therapies for AaD will assume greater significance than new highways or higher salaries for police officers. The dedication of sufficient financial resources leaves the problem of finding healthy human test subjects. There is an obvious place to find them.

Those who truly want safe and effective therapies of AaD but are unwilling to join the ranks of human test subjects required to test them will do what the relatively wealthy have always done: they will pay others to do their dirty work for them. Rich people pay poor people to clean their swimming pools. They will have the option, either as individuals, or as members of consortia, to pay the poor to serve as guinea pigs for SENS. The poor world contains many millions of healthy people who would be, in clinical terms, perfect subjects for SENS experiments. These perfect subjects will be experiencing what we might call healthy aging. A wealthy supporter of SENS can feel reassured that a therapy shown to be both safe and efficacious against AaD in a clinical trial involving hundreds of such subjects will work for her. We can imagine that many poor people would gladly sign up to participate in a trial for some new AaD therapy. They may have no rational expectation of benefiting from the trial. They may expect it to shorten their lives. But monies paid to them address other significant and proven obstacles to human flourishing. Money for participating in SENS trials will provide adequate shelter, nutrition, access to basic health care, and so on. Those who die in a trial that shows that one way to remove the telomerase gene from every cell in the human body does not work will provide educations and medical treatments for their children. Poor people will compare the possible or probable dangers of participating in a clinical trial for an experimental AaD therapy with the established dangers of poverty. Exactly the same logic would see the rich world buying the children of the poor, the kidneys of the poor, and the kidneys of the children of the poor.

The medical exploitation of the poor world is tellingly depicted in John Le Carré's *The Constant Gardener*. In this novel, a Western pharmaceutical company manipulates Kenyans into clinical trials of a dangerous experimental tuberculosis drug that would not be approved for clinical trials in any rich country. I suspect that anti-aging therapies pose a greater threat than Le Carré's imaginary TB medicine. The sums of money to be realized from a clinically proven agent of rejuvenation are surely much greater than those that would flow from a better drug for tuberculosis. The intended market for SENS medicines are conventionally healthy rich people and not sick poor people. The incentives for unscrupulous tests of experimental therapies for AaD will be correspondingly greater than those that Le Carré imagines motivating the fraudulent TB trial.

There have been moves to restrict this exploitation. In late 2011, the United States Presidential Commission for the Study of Bioethical Issues recommended a series of recommendations designed to offer stronger protections for participants in clinical trials. President Obama suggested that the commission give special consideration to an injustice done by medical researchers to the people of Guatemala in the 1940s. Vulnerable people were deliberately infected with sexually transmitted diseases by US researchers. The commission sought to reduce the likelihood of this kind of injustice by demanding "equivalent-protections." American companies and researchers conducting research in the poor world would have to grant patient protections that are the same or better than those protecting participants in domestic trials.

I am pessimistic about the prospect for such restrictions when the citizens of rich world nations get a genuine sniff of anti-aging treatments that truly address the fulcrum of aging. One factor limiting the roll out of efficacious SENS therapies is available funds. But another factor is the lack of live, healthy bodies for SENS clinical trials. I've suggested that the poor world is an obvious place to recruit human test subjects. The years of the late twentieth and early twenty-first centuries are a historical anomaly in relations between the rich world and the poor world. In the past the rich have viewed the poor as having something of value. They could be enslaved and their lands taken. In the current age, the rich are effectively prevented from enslaving the citizens of the poor world. The most obvious and potentially lucrative forms of colonization—the settling of their lands and extraction of their resources—are prevented. The chief contemporary problem is one of indifference. The poor find themselves excluded from the global economy. I suspect that the end of the pro-aging trance will reintroduce the poor to the global economy. Those among the poor who are free of disease will be required for clinical trials of SENS therapies. Unlike slavery, this is a form of exploitation that happens with the consent of the exploited. Few people volunteer to be enslaved. But people effectively excluded from the global economy face a traumatizing choice. They are likely to see selling themselves as guinea pigs for SENS as an opportunity to connect with that economy.

Will Volunteer Risk Pioneers Help Out?

Perhaps I'm unduly pessimistic about the prospect of attracting sufficiently many people enjoying a healthy middle age to serve as the human test animals for SENS therapies, therapies that if they are to work as intended must make quite fundamental revisions to the aspects of human biology that permit aging. When I presented an early version of these ideas in *Slate* magazine, the libertarian transhumanist writer Ronald Bailey responded to me.[12] I had speculated about what might be involved in arranging a clinical trial of WILT. It seemed unlikely that the kinds of experimental evidence that might be gathered in support of WILT could make it rational for a healthy person to participate in a clinical trial of it.

Perhaps I'm exaggerating. Bailey expresses enthusiasm for experimental work on mice that seems to support a hypothesis about the relationship between telomeres and aging that differs from the one behind WILT. The telomeres of our cells tend to get shorter with each division. Eventually they reach a state at which they are less able to protect the cell's DNA from damage. According to one line of reasoning, if telomeres could be returned to their youthful, long state, they could better protect DNA from damage. The secret of eternal youth might, on this reasoning, come in the form of a therapy that lengthened telomeres. Bailey points to the people who pay for experimental drugs that may lengthen their telomeres in the hope that this will lengthen their lives. He says, "some rich people who are eager to extend their lives are already paying considerable sums to take a supplement, TA-65, which has been shown to lengthen telomeres in humans."[13] Bailey rejects my suggestion that the poor will be targeted for testing SENS theories. According to him, "No poor people need apply." There's manifestly no need to coerce the poor into doing what the rich not only do voluntarily, but are prepared to pay large sums of money to do.

This general attitude that Bailey describes receives endorsement from Allen Buchanan. According to Buchanan, the rich may perform a valuable service through their earlier access to a biomedical enhancement. Wealthy people play the role of "volunteer risk pioneers."[14] He says:

They perform a valuable service: They buy the first generation version of the innovation at a high price. Often they get a defective product and sometimes a dangerous one. Later, when the bugs have been eliminated and the safety issues have been dealt with, you and I buy the improved version at a lower price. How's that for a deal? And we don't even have to force them to do it.[15]

I suspect that wealthy people will make a distinction between experimental technologies such as cell phones and computers and experimental modifications to their brains and bodies. There's a difference between a 1980s experimental mobile phone that occasionally drops a call and a defective experimental therapy that shortens your life.

It's important not to underestimate the requirement of SENS for volunteer risk pioneers. Some fields of inquiry can get by with relatively few risk pioneers. The Wright brothers accepted risks, successfully flew, and showed that powered flight was a possibility for human beings. In the case of SENS, the self-sacrifice of a few mavericks is unlikely to suffice. SENS requires that we recruit sufficient numbers of volunteer risk pioneers to satisfy the testing requirements of clinical trials on seven independent lines of research on the fundamental causes of aging.

I wonder about the rationality of volunteer risk pioneers eagerly popping TA-65. The company cites evidence that lengthening of telomeres improves the "immune system, vision, male sexual performance, skin appearance, and more."[16] The website of T.A. Sciences, the company selling TA-65, carries the reassuring message that "TA-65® has been in use since 2005 with not one reported adverse event."[17] How happy should T.A. Sciences' clients feel about this statistic in light of de Grey's hypothesis about the therapeutic value of doing the opposite of what TA-65 is supposed to do? De Grey hopes to rejuvenate us by entirely eliminating telomerase from human bodies. This should decisively prevent the lengthening of telomeres. At this stage in our understanding, there is some support for de Grey's hypothesis. The gene that makes telomerase exists in almost every cell in the human body. For some reason it is switched off in all but a few cells. Could this be an accident? De Grey thinks that this is an evolved response to cancer. The gene is active only in cells that are, of necessity, frequently copied—cells that make blood and replenish the stomach lining. If de Grey's hypothesis is correct, then T.A. Sciences' customers might pay for their experience as life extension risk pioneers with an elevated risk of cancer. The elevated levels of telomerase will help incipient tumors to survive until additional mutations enable them to decisively evade the body's defenses against cancer. Perhaps other would-be life extenders will be grateful that they have chosen to leave medically useful corpses when they could more enjoyably have chosen to elevate their cancer risk by smoking cigarettes. Of course, de Grey's views about the dangers of the telomerase gene may be false. There could be some entirely innocent reason for its unexpectedly low level of activity in the human body. But the point is that it may take quite a long time to detect an elevated risk of cancer from TA-65.

Suppose that de Grey is wrong about the relationship between cancer and the shortening of telomeres and T.A. Sciences is right. We should ask about the expected scale of TA-65's effect on our bodies. De Grey's approach treats aging as a by-product of the fundamental workings of human bodies. This is why SENS promises so much. According to T.A. Sciences, "TA-65 is a natural molecule derived from the Astragalus plant, a Chinese herb used since ancient times."[18] Could an herb in use since ancient times really address the, or a, fulcrum of aging? It seems to be a more superficial intervention than are many therapies that de Grey anticipates making up SENS. TA-65 is likely to be significantly less dangerous than any of the therapies that interest de Grey, precisely because of its likely lack of effect on any fundamental causes of aging.

There may be opportunities for volunteer risk pioneers to more effectively contribute to SENS without subjecting themselves to unwanted risk. SENS is a big, ambitious program comprising many hypotheses about the causes of human aging and ways to fix them. This point notwithstanding, I think that there is a general reason that SENS is unlikely to attract many volunteer risk pioneers. The ideal that motivates it is intrinsically less disposed to inspire acts of self-sacrifice. When Thomas Jefferson said, "The tree of liberty must be refreshed from time to time with the blood of patriots and tyrants," he was presenting liberty as an ideal sufficiently worthy to motivate self-sacrifice. Science boasts many examples of ideals sufficiently worthy to motivate self-sacrifice. Explorers risked their lives to expand our knowledge of the world. Chemists voluntarily placed themselves into close physical contact with toxic elements such as mercury and radium, shortening their lives in the process. These risky acts were motivated by a powerful ideal—the advance of our understanding of the world. Medical researchers place themselves in close contact with infectious agents because they are motivated by a desire to end the suffering that these agents cause. We should not overlook the tradition of self-experimentation in medicine. A notable recent example is the self-experiment performed in 1984 by the Australian Nobel laureate Barry Marshall on the causes of peptic ulcers. The received view at that time was that the vast majority of peptic ulcers resulted from stress, spicy food, or an overabundance of gastric acid. Marshall swallowed samples of the bacterium *Helicobacter pylori*, gave himself gastritis—a recognized cause of peptic ulcers—and cured himself with antibiotics. Marshall was a scientist with a passionate desire to find the truth. He was prepared to make himself sick to find it. Might SENS inspire similar acts of self-sacrifice?

I doubt that it will. Those who should be eligible to be volunteer risk pioneers for SENS face a practical dilemma. We can say that they are either interested in SENS and desire to see it progress, or they are not. Those who are not interested in SENS are likely to find other outlets for their risk pioneering. Those who are interested in SENS and want to see it progress are likely to be especially averse to the specific kinds of sacrifice its form of risk pioneering demands. They are being asked to put at risk precisely the thing—the possibility of a long life span—that commitment to SENS assumes.

Consider an analogous case. Suppose that someone devises a plan to make some millionaires even richer than they currently are. The success of this plan requires a small number of people to make a sacrifice. They are not expected to risk their lives. Rather, they are expected to risk their retirement savings. To create even greater wealth for the millionaires, they must place their own savings at risk. Suppose that there is no prospect of any compensation from the even richer millionaires. I suspect that it will be very difficult to persuade people to make this sacrifice. Some people won't be interested because they don't think that making rich people even richer is a particularly important thing to do. But those who do think that the world would be a better place if Donald Trump and his ilk became much richer are likely to be deterred by the specific form of the sacrifice. They are asked to put at risk precisely the thing—wealth—that they are, in practical terms, required to view as very important in order to be at all interested in the project. They are required to risk their wealth so that others who are already wealthy can become even wealthier. This is analogous with the choice that confronts risk pioneers who express an interest in SENS. Promoting the cause of SENS requires risking precisely the thing—one's health—that they must be interested in to want to sign up in the first place.

Ethical Anti-Aging Experiments Not Now, but Someday?

One way to achieve effective safe therapies for AaD requires that sufficiently many people be inspired by the prospect of agelessness to sign up for clinical trials. I think that this is unlikely to happen. Alternatively, the healthy poor could be induced to sign up. I think this is somewhat likely to occur. And it would be an exceedingly bad thing.

There is another option: simply to wait. A historical relativity is implied by the term "normal" in "normal human life expectancy" and "normal human life span." Our views about what is a normal human entitlement have changed quite significantly over the last centuries. We now think of a life span of three score years and ten plus some as a normal human entitlement. In the past, people's expectations were more meager. A twelfth-century European peasant might have expected a life span of around three decades. Suppose that over the next decades we make steady progress against recognized DoAs. We find therapies that keep in check the advance of Alzheimer's. We make steady progress against heart disease. Life expectancies will, somewhat gradually, get longer. This gradual rate of improvement will lead to a change in popular understanding of what is normal. There may come a time when people who expect to achieve a life span of only eighty years will feel that they are missing out on the century to which they are entitled. A series of successes against DoAs—Alzheimer's, heart disease, diabetes, and the like—should shift attention to AaD. We might then see a rational interest in participating in clinical trials for SENS therapies.

The point is that this won't happen overnight. Medical research should continue to focus on the problem of how to give people with Alzheimer's, heart disease, and other recognized DoAs additional good quality years. It's only when these difficult problems are effectively solved that the focus of the medical establishment should shift to AaD.

7   A Defense of Truly Human Enhancement

The chief focus of the book so far has been on the dangers of too much enhancement. The intrinsic value of human enhancement conforms to an anthropocentric ideal. Beyond a certain point, greater degrees of enhancement sever the connection with internal goods and therefore reduce the intrinsic value of enhanced capacities. Note that this is a statement about our present values. It is not intended as a statement about the values of beings we could possibly become. We could undergo a transformative change of radical enhancement that would exchange one way of evaluating our experiences and achievements for another. This should no sooner lead us to endorse radical enhancement than anticipation of the psychological effects of body-snatching or cyberconversion should lead us to value those transformative changes. It is prudent to privilege our current evaluative framework over others that enhancement might give us. In addition, radical enhancement threatens the autobiographical memories that underlie our sense of ourselves as beings persisting through time.

In this chapter, I switch our focus to enhancements of a lesser degree. I argue that failures of imagination and identification that prevent us from fully valuing radically enhanced experiences and achievements do not prevent us from valuing experiences and achievements of a lesser degree. Lesser degrees of enhancement need not threaten the continuities of autobiographical memory that underpin our identities. The experiences enabled by lesser degrees of human enhancement may be prudentially valuable. In chapter 1, I introduced a distinction between radical enhancements—enhancements that improve significant attributes and abilities to levels that *greatly exceed* what is currently possible for human beings—and moderate enhancements—enhancements that improve significant attributes and abilities to levels *within or close to* what is currently possible for human beings. The ideal of truly human enhancement defended in this book endorses some moderate enhancements while rejecting all radical enhancements. It locates itself between two more extreme philosophical ideals about enhancement. Bioconservatives, including the public intellectual Leon Kass, the political scientist and historian of ideas Francis Fukuyama, and the environmentalist Bill McKibben, reject any manner of human enhancement. Transhumanists, including the philosophers Nick Bostrom and Mark Walker, the sociologist and ethicist James Hughes, and the cryonics advocate Max More, endorse very many forms of human enhancement. They reject the moral or prudential significance of limits conditioned by our humanity and so tend to endorse an objective ideal for enhancement. According to the transhumanists, more is almost always better.

In this chapter, I argue indirectly for moderate human enhancement. I argue for a defeasible presumption in favor of moderate enhancement. Some degree of human enhancement is widely recognized as good practice in education and nutrition. For example, educators do not recognize a moral limit fixed by species-norms in their attempts to make our children more knowledgeable. The presumption in favor of human enhancement could be defeated by sufficiently powerful reasons. Perhaps some forms of human enhancement are inimical to human flourishing. The dominant strand of the anti-enhancement literature focuses on the means by which humans are enhanced. It identifies as especially problematic attempts to enhance humans by selecting or modifying human genes—it opposes *genetic* enhancement. This chapter explores and rejects six ways in which genetic enhancement is supposed to be worse than enhancement achieved by modifying environmental influences. We are left with the idea that morally and prudentially good ways of enhancing humans include some genetic enhancements.

I hope that my arguments against too great a degree of enhancement will make genetic enhancement seem less frightening. George Annas would like to see an international human species preservation treaty that would protect against species-endangering acts that include genetic enhancement.[1] The fact that an enhancement is genetic does not make it dangerous to the human species. Of greater relevance is the degree of the enhancement. If they are to be drawn up, species preservation treaties should focus on degrees rather than on means of enhancement.

The Ubiquity of Human Enhancement

In chapter 2, I suggested that we can better understand some of the ethical disputes over enhancement if we acknowledge two concepts of human enhancement. There is enhancement *as improvement*. On this account, any time we improve a human being we achieve human enhancement. To suppose that opponents of human enhancement oppose any improvement of human beings is to foist on them an untenable position. The concerns of opponents of human enhancement make more sense if we interpret them as objecting to enhancement *beyond human norms*. This definition of enhancement makes reference to biological norms. It contrasts enhancement with therapy that includes measures designed to restore or preserve normal levels of biological functioning. When given to someone with anemia, synthetic erythropoietin (EPO) is therapy. When given to healthy, fit, Tour de France cyclists, it enhances beyond human norms. In this chapter, I understand opponents of human enhancement as objecting to deliberate enhancement beyond human norms.

There is more than one way in which we can attach philosophical significance to the distinction between therapy and enhancement beyond human norms. Bioconservative thinkers tend to use it to mark the difference between *permissible* and *impermissible* ways of making or changing human beings. For example, Francis Fukuyama allows that therapeutic interventions are permissible. He argues that interventions whose purpose is to enhance are impermissible.

In what follows, I show that this use of the therapy–enhancement distinction is unsustainable. This is not to say that the distinction does not indicate anything of moral significance. Consider, for example, the use of the distinction made by Allen Buchanan, Dan Brock, Norman Daniels, and Daniel Wikler in their book *From Chance to Choice*. According to Buchanan and his coauthors, the category of therapy corresponds approximately to those genetic interventions that the liberal state should seek to provide to its citizens, supposing that the resources exist. Liberal states may be subject to no obligation to commit resources to make morally permissible enhancements available to their citizens—enhancements that we do not have reason to oppose can be left to the discretion of individuals or their parents. Enhancements, on this view, may properly be likened to after-school tuition in a musical instrument. The state should ensure that good quality educational services are available to its citizens. It can leave the provision of oboe lessons to those with an interest in teaching the instrument and parents with an interest in purchasing oboe lessons for their children. Rather than indicating a difference between permissible and impermissible modifications, the therapy–enhancement distinction marks the difference between permissible modifications that the state should provide and permissible modifications whose provision it can leave to individuals.

I suspect that the therapy–enhancement distinction is much better suited to the role envisaged by Buchanan and his coauthors than it is to the role imagined by bioconservatives. Buchanan and his coauthors present therapeutic genetic modifications as—in general—helping to provide citizens with the prerequisites for normal participation in society. The distinction between therapy and enhancement is, for them, a politically useful approximation to the complex collection of capacities that enable normal participation. The preservation or restoration of normal biological functioning serves as a focal point for a variety of interests in protecting equality of opportunity. For example, certain variations in human DNA are associated with degrees of cognitive disability that prevent normal participation. Gene fixes would eliminate this barrier. Those who lack basic literacy and numeracy find their options for meaningful participation in early twenty-first-century liberal democratic societies severely restricted. There is, of course, a significant degree of relativity in such assessments. Illiteracy in a preliterate hunter-gatherer society is no handicap to normal participation in that society. Buchanan and his cowriters imagine that in future societies certain enhancements may be required for normal participation. "Suppose we have a genetic technology that allowed us to enhance immune capabilities beyond those involved in normal functioning. Then, like vaccinations—which have an analogous effect—we might well be obliged to provide this enhancement as part of a medical benefits package (costs and resource constraints permitting)."[2] They go on to speculate about inclusion of a variety of cognitive enhancements in this collection of genetic interventions that the state should provide to its citizens.

I will have nothing more to say about use by Buchanan and his coauthors of the therapy–enhancement distinction. In what follows, I show that the distinction is deeply implausible when used to mark the difference between permissible and impermissible interventions.

As stated above, the distinction between therapy and enhancement makes no reference to the means by which either category of intervention occurs. For example, an enhancement may be produced by the modification of genetic material. It may result from the alteration of some aspect of our environments—would-be enhancers of humans may seek to discover a dietary supplement that prolongs life or discover a new technique for teaching calculus. It is deeply implausible to seek to prevent environmental enhancement. The ambitions of educationalists looking for new techniques to teach algebra and nutritionists exploring ways to improve human diets are not and should not be limited by human norms.

Suppose there is a risk-free educational technique capable of improving the scores of school children by 5 percent on some widely accepted test of mathematical aptitude.[3] The effects of this new technique are uniform across the human population: it adds 5 percent to the scores of all students to whom it is offered. The distinction between therapy and enhancement reveals differences in the technique's effects. Toward the bottom of the ability spectrum it may be viewed as performing a remedial role. Here the technique would be a kind of educational therapy, bringing children who start with a deficit in mathematical understanding closer to achieving results that, before the advent of the new technique, would have placed them in the normal range of human achievement. Children toward the upper end of the ability spectrum already exceed the levels of attainment that educationalists use to define normal mathematical understanding. By pushing them from a point above the normal range to a point further above the normal range, the new technique enhances them.

It would be absurd to use the suggestion that when presented to some children the technique's effects are therapeutic but when presented to others it is an enhancement to justify providing it to the first group but making it unavailable to students in the more talented group. The state might view educational therapy as a more important goal than educational enhancement. If the technique is very expensive it might decide to prioritize the struggling students. But this claim about government spending priorities does not support banning provision of the technique to more talented students. If the parents of mathematically talented students want to pay for the technique, they should not be prevented from doing so.

It seems that the idea that the distinction between therapy and enhancement marks the distinction between permissible and impermissible interventions can be taken seriously only if it is restricted to *genetic* therapies and enhancement. In the remainder of this chapter, I argue against such a restriction. Our dominant mode of evaluating enhancements should focus not on the means by which they are produced—whether they result from the modification of genes or environments—but instead in terms of the degree to which they enhance.

Enhancement and Heredity

The idea of enhancing humans by selecting or modifying human hereditary material has an ugly history. The first systematic expression of this idea of human improvement by the selection or manipulation of hereditary material was eugenics. The word "eugenics" first appears in Francis Galton's 1883 work *Inquiries into Human Faculty and Its Development*. Galton was a cousin of Charles Darwin who was among the first to explore the social implications of the theory of evolution. It combines the Greek *eu*, meaning "good" or "well" with the suffix *-genēs*, meaning "born." Galton defined eugenics as "the science of improving stock, which is by no means confined to questions of judicious mating, but which, especially in the case of man, takes cognizance of all influences that tend, in however remote a degree, to give to the more suitable races or strains of blood a better chance of prevailing speedily over the less suitable."[4]

A great deal of evil was done in the name of eugenics. While the Nazis' attempts to defend certain of their murderous policies as eugenics were certainly the moral low point, they were not the only offenders. Galton's science of improving human stock motivated and was thought to justify marginalization and involuntary sterilization throughout Europe, North America, South America, Asia, and Australia.

That eugenics has a horrible history cannot be denied.[5] But eugenics' historical failings should not taint a modern program of enhancing humans by selecting or manipulating hereditary material if that program can be shown to be different in the morally significant respects. There are many differences between Hitler's eugenics and a program of genetic enhancement that might be acceptable to the citizens of a culturally diverse, early twenty-first-century, liberal democracy. The former was driven by a hodge-podge of unscientific prejudices; the latter would be informed by the understanding of human heredity brought by modern genetics.[6] The former imposed a monolithic view about human flourishing; the latter would acknowledge a plurality of views about the good life. The former used murder as a tool of human improvement; the latter would insist on strict moral limits on individuals' enhancement plans.

This chapter presents and rejects six ways in which genetic enhancement might be morally worse than environmental enhancement. Moderate enhancements do not cause the failures in engagement caused by radical enhancement. Suppose that we do agree that the main danger stems from enhancements of too great a degree. I make a case that the main threats from enhancement are likely to be environmental rather than genetic.

Defining Genetic Enhancement

There are a variety of techniques by which humans could be genetically enhanced. Some of these techniques involve the selection of human genetic material. Others involve its modification. On the selection side is pre-implantation genetic diagnosis (PGD). Couples undergoing PGD have a number of embryos created by in vitro fertilization (IVF). These embryos are grown to the eight-cell stage whereupon a single cell is extracted and the DNA of that cell is analyzed. The results of this analysis determine whether or not the embryo is used to arrange a pregnancy or discarded. PGD is used primarily to avoid serious diseases. Among these diseases are the neurodegenerative disorder Huntington's disease, cystic fibrosis, and a range of other diseases. In most jurisdictions, PGD is not licensed for the purposes of boosting performance beyond human norms. But there is nothing in human biology or the technique itself that would prevent it from being used for these purposes. PGD's effectiveness as a tool of enhancement is limited by the numbers of embryos that can be created by IVF. It's likely that at least some of the causes of Albert Einstein's exceptional scientific achievements resided in his genome. Einstein's particular combination of genes arose by chance. Parents interested in having a baby Einstein can attempt to improve their chances by using PGD. First they would need to identify specific genetic variants that contributed to Einstein's scientific genius. They could search for these variants in embryos created for them by IVF. They would improve their odds of finding the combinations that they seek by increasing the number of embryos that they create by IVF. The specific Einsteinian combinations may be unlikely to occur in any embryos that a couple could jointly produce. Those who seek to have a baby Einstein will probably have to content themselves with embryos that carry some among the many genetic variants deemed to have contributed toward Einstein's talents.

Even if successful, there would be no guarantee that someone brought into existence in this fashion would achieve scientific genius. Perhaps he or she would be a feckless computer hacker or a sporadically employed laborer. But if we've identified the hereditary factors relevant to Einstein's achievements, then a child whose genome bears these factors should at least be viewed as having a head start on the way to success as a scientist.

PGD selects from among existing variation. Genetic engineering is a seemingly more powerful technique that involves the modification of a genome. To elaborate on the above example, suppose that the genetic variants that contributed toward Einstein's genius were identified. These might be introduced into the genome of a developing child. If this introduction occurs sufficiently early in the process of human development, then the Einsteinian variants would be transmitted to every cell in the developing child. As above, there are no guarantees of scientific genius. All that can be claimed is that the resulting child might have a somewhat increased likelihood of being cognitively gifted.

The Interactionist View of Development

The notion that environmental interventions resulting in enhancement should be treated differently from genetic interventions that have the same effect seems to be in tension with modern understanding of development. This *interactionist view* is opposed both to genetic determinism according to which human beings are shaped almost exclusively by genes, and environmental determinism that says that we are made by our cultures, educations, diets, and a host of other environmental factors. According to interactionism, we result from the complex interaction of tens of thousands of genes and uncountable environmental influences. There's no point in venturing guesses about which category of influence is more important. Genes acting alone cannot make a human being. But nor can environments. Identifying yourself with your genetic material is as ludicrous as literally, rather than metaphorically, identifying yourself with your school's educational philosophy. British Prime Minister David Cameron is "an Eton man." This is a claim about where he went to school—it does not purport to capture an indispensable component of his identity. Both our environments and our genes stand at least one causal step away from the properties—our memories, conscious thoughts, and so on—that constitute us. They are partial *causes* of the characteristics that make us who we are. They are not identical to them.

This assertion of the parity of genetic and environmental influences may seem to be at odds with the oft-presented message that certain human traits are largely genetic while others are largely environmental. For example, human eye color is often presented as a genetic trait. Height is, in contrast, partially genetic and partially environmental. Any apparent contradiction resolves after clearly separating two questions about the relationship between genes and environment. The statement that eye color is principally genetic addresses variation in human populations. It's the claim that most of the observed variation in eye color in human populations corresponds with variation in genes. When we compare groups of humans with brown eyes with groups of those with blue eyes, the relevant differences lie largely in DNA.

A focus on the development of individuals exposes a different relationship between genes and environment. Here we're not interested so much in variation in current human populations, but instead in a particular developmental story—the specific causal contributions that genes and environment make toward a given individual's traits. I have brown eyes. Suppose the tape of my life were to be rewound to conception and restarted. It's likely that certain changes to my environment—the introduction of certain viruses or specific dietary modifications—would change my eye color. The sum of these counterfactual changes corresponds with the parts of my environment that are causally relevant to my eyes' brownness.

Parity is not identity. To say that genes aren't more developmentally significant than the environment, and vice versa, is not to say that they make identical contributions. Their contributions are very different, and these differences must be understood by anyone hoping to enhance human performance. A given culture's understanding of human development and how to manipulate it determines what enhancements are available to its members. For example, it's relatively easy to imagine Roger Federer discovering a new training regime—an environmental modification—capable of improving his already excellent tennis forehand. We have, in contrast, few clues about any genetic modifications that might produce the same result. This is certainly not to say that we won't discover them at some point in our future. It's possible that there are yet-to-be discovered genetic variants that, once introduced into Federer's genome, would enhance his visual system, allowing speedier analysis of the angle and velocity of balls coming to his forehand. At the very least, their possibility should not be ruled out in advance of investigation.

Six Ways in Which Genetic Enhancements Could Turn Out to Be More Morally Problematic Than Environmental Enhancements (but, in Fact, Do Not)

Interactionism leads to the rejection of in-principle moral differences between genetic and environmental enhancements. The lack of a principled difference leaves plenty of room for differences in practice. Consider a genetic enhancement and an environmental enhancement that resemble each other in all respects except for how they are produced. One results from modifying a genetic influence, whereas the other results from modifying an environmental influence. We may be rationally required to draw the same conclusion about their morality. This still leaves open the possibility that, considered as a class, genetic modifications may in fact have different effects and uses from the class of environmental modifications. In what follows, I explore six ways in which genetic enhancements might turn out to be more morally problematic than environmental enhancements.

(1) Are genetic enhancements more morally problematic because they are of greater magnitude than environmental enhancements?

Perhaps genetic enhancements belong in a different moral category from environmental enhancements because their effects are (always or potentially) of greater magnitude. We know that there are environmental modifications—improvements to diet and education—that influence human intelligence. But there are limits on how smart such modifications can make us. Even the most enthusiastic advocates of omega three think that a diet rich in oily fish can boost intelligence only to a modest extent. It's not a way to turn an average achiever into a genius. In popular presentations, at least, limits such as these appear not to constrain genetic engineers who are free to supplement existing intelligence genes with additional copies or to invent intelligence genes with novel modes of action.

Some of the apparent power of genetic enhancement comes from the fact that the chief venue for its presentation is science fiction. Suppose your story's central character has intellectual powers beyond those of any past or present human being. Readers will more readily attribute this trait to the unknown future (and therefore more sci fi) technology of genetic enhancement than they will to measures with which we are familiar, such as diets rich in oily fish or alfalfa.

It's possible that genetic enhancement could produce significant increases in intelligence. At least, it's dangerous for moral philosophers to proceed on the assumption that they won't. But it's also far from true that a few millennia of experiments in educating humans have exhausted ways in which human intelligence might be boosted. For example, the psychologist Anders Ericsson has done a great deal of work on the acquisition of skills by "deliberate practice."[7] Deliberate practice is more than just practicing hard—it involves sequences of activities carefully selected with the purpose of extending skills. Done right, deliberate practice seems capable of producing quite remarkable results. Ericsson cites the case of the Hungarian child psychologists László and Klara Polgár, whose program of extended deliberate practice—tens of thousands of hours of learning from failures and not resting content with successes—turned their three daughters into some of the world's strongest chess players. The Polgárs' particular way of implementing extended deliberate practice satisfies our definition of enhancement—the sisters' chess talents were well beyond human norms. László and Klara certainly didn't cease their program of deliberate practice when their daughters achieved a level of competence in chess that might be considered normal for humans.

In some of the more radical literature on our species' future, genetic modification is almost old hat. For example, Ray Kurzweil advocates grafting a variety of cybernetic implants and neuroprostheses to our bodies and brains. He imagines electronic neuroprostheses that will dramatically enhance our mental powers. Kurzweil predicts a gradual merger of human with machine. In its early stages, this merger will be motivated by a desire to fix parts of our brains that have become diseased. Cochlear implants already help profoundly deaf people to hear by directly stimulating their auditory nerves. Soon prosthetic hippocampuses could be restoring the memories of people with Alzheimer's disease. Once we install the implants, we will face a choice about how to program them. We hope that they can at least match the performance of the parts of the brain they replace; we hope, for example, that prosthetic hippocampuses will be as good at making and retrieving memories as healthy biological human hippocampuses. But if you've gone to all the trouble of installing a prosthetic hippocampus, why would you rest content with a human level of performance when you could have so much more? From a technological perspective, there's nothing sacred or special about our present intellectual powers. This attitude to the machinery of thought will lead, in the end, to a complete mechanization of the human mind. Kurzweil presents the resulting massively intelligent machine minds as, at one and the same time, completely nonbiological and fully human. The procedures that will introduce these devices into human brains do not modify genes. Their mode of enhancement is environmental.

To summarize, there's no reason to believe that environmental and genetic enhancements must differ in degree. Of course, a particular genetic modification may enhance a trait such as intelligence to a greater extent than a particular environmental modification. But in such cases it's the extent to which intelligence is enhanced that matters rather than the means by which this is achieved.

(2) Are genetic enhancements more morally problematic than environmental enhancements because they pose a greater threat to our humanity?

We know that environmental modifications can produce all sorts of weird and wonderful effects on humans, ranging from extended directed practice in chess to full body tattoos. Strange though these modifications are, their subjects remain recognizably human. No one denies the humanity of the Polgár sisters or of Erik Sprague, a.k.a. the Lizardman, who has pursued his own personal enhancement agenda, tattooing his entire body with green scales, splitting his tongue, and making plans to acquire a tail transplant. Modifying human DNA differs in potentially pushing us beyond the genetic boundaries of the human species. Perhaps it will transform us into a new biological species of posthumans.

The view of species as having definite genetic boundaries has come under threat from recent work on the relationship between development and evolution. According to the picture emerging from evolutionary developmental biology (the so-called evo-devo approach), the novel traits of a new species do not emerge from the invention of a host of new genes. They result largely from alterations to the regulation of elements in a shared genetic tool kit. On this view, birds evolved from theropod dinosaurs through modifications of switches regulating the expression of genes which themselves have been conserved. Birds didn't lose tail genes and gain wing genes. They lost their tails and gained wings by differently regulating the genes that in theropods both produced tails and directed the development of arm bones. The paleontologist John Horner presents a vision of a Jurassic Park in which Tyrannosaurs are restored to life not by the cloning of preserved T-Rex cells, but instead by altering the timing of the relevant genes in the chicken genome.[8] On the evo-devo view, it's possible that a human genome could yield something that was manifestly nonhuman by altering the timing of a few key genes. This could be achieved by the insertion of new genetic regulatory sequences. But the fact that many regulatory signals originate from the environment opens the possibility of growing a posthuman from a human embryo by manipulating its early environment so as to achieve a different sequence of gene expression.

The possibility of cybernetic enhancements provides conceptually clearer cases of genetically human nonhumans. It's wrong to think that because they don't modify genes they leave their subjects' humanity intact. The (mostly) malevolent cyborgs that give their names to the *Terminator* movies are constructed by grafting a human epidermis over a machine body. To the extent that these cyborgs are genetically anything, they're genetically human. Yet it seems mistaken to think of them as human. The cyborgs are more properly viewed as genetically human nonhumans.

Later I present an argument for the claim that certain enhancements do in fact threaten our membership of the human species and, by implication, our connection with distinctive human values. This threat does not arise from the means by which these enhancements are engineered. There's no reason to think that genetic modification is intrinsically more likely to turn us into posthumans or into any other kind of nonhuman than is environmental modification.

(3) Is genetic enhancement worse than environmental enhancement because it's less natural?

There is an often-made claim that genetic modification or selection is an unnatural means of enhancing human capacities. It involves deliberate interventions in our natures that have until very recently been impossible for us. It is therefore more suspect than environmental means of enhancement that use the traditional and therefore more natural modes of enhancement of better diets or better lessons.

Such moral or prudential appeals to nature are highly suspect. Reading glasses are an unnatural but morally and prudentially impeachable aspect of modern existence. The naturalness of senile dementia offers no rational support for its retention. But the following paragraphs concede, for argument's sake, that naturalness could offer some manner of moral or prudential endorsement.

I propose that genetic enhancement should be acknowledged as a comparatively natural way of enhancing humans. What do we mean when we say that a technology is natural? No technology can be entirely natural—the word "natural" is defined so as to exclude human activity. But there are usages of "natural" that admit of degrees. Wholemeal bread is more natural than white bread because the degree of processing it undergoes is less. In wholemeal bread, entire grains are present. In the white bread, these grains are broken down. Reference to degrees of processing can explain why we are right to call wind power a more natural way of generating power than burning coal or nuclear fission. The generation of energy by the use of a wind turbine is a human technology. It cannot therefore be entirely natural. But it makes direct use of a natural process. Coal-fired plants require coal to be dug up from deep underground, transported to power plants, and then burned to generate the heat necessary to turn turbines. They are, as a consequence, less natural than wind turbines.

We can acknowledge degrees of naturalness in the manners in which humans are enhanced. Genetic enhancement is more natural than many forms of enhancement in that it makes use of natural processes. The NR2B gene plays a role in making brain tissue appropriately connective. If the experiments on mice are any indication, the insertion of an additional copy of the NR2B gene could make humans more intelligent by increasing the connectivity of brain tissue. This form of genetic enhancement exploits a natural process corresponding with the role NR2B plays in normal human development. Genetic enhancers effectively co-opt natural design. Genetic enhancement may be less natural than extended deliberate practice. But it is more natural than the forms of enhancement that Kurzweil anticipates will supersede it.

(4) Is genetic enhancement less fair than environmental enhancement?

If genetic enhancements follow the pattern set by other new technologies, they are likely to be very expensive when first introduced and therefore available only to the wealthiest among us. The rich will, as a consequence, supplement their existing environmental advantages with genetic ones.

This concern should not be minimized. But it attaches both to environmental enhancement and to genetic enhancement. There's no reason to predict that a super-efficient electronic hippocampus capable of enhancing human memory will be cheaper than a genetic enhancement with this effect.

Concerns about fairness apply to techniques of environmental enhancement that exist now. Take the home-schooling program that turned the Polgár girls into chess masters. Now consider the same technique directed at attributes more directly connected with economic success. Imagine a program of extended directed practice constructed on the basis of what was learned about the financial system in the wake of the 2008 financial crisis. Children schooled this way might acquire the capacity to identify stocks and bonds that should be sold short that would be mistakenly described by observers as signs of inborn financial genius. One could imagine that the very wealthy who find their time too valuable to home-school their kids will have the option of hiring experts to give their children 10,000 hours of directed practice at manipulating financial markets, an option unavailable to the poor.[9] There's a danger that the rich will supplement their already existing environmental advantages with even more powerful, very expensive environmental enhancements.

The factors that rightly prompt concerns about fairness and price and other similar barriers to access are not specifically genetic or environmental. They arise in connection with the broader category of enhancement. Presenting unfairness as a consequence of specifically genetic enhancements leaves us less able to anticipate and respond to inequalities and injustices brought by environmental enhancements.

(5) Are genetic enhancements more morally problematic than environmental enhancements because they tend to conflict with the recipient's autonomy?

The best opportunities to genetically enhance an organism arise at the very beginning of that organism's life. A gene successfully introduced into a single cell human embryo should be transported, by successive cell divisions, into every cell in the body of the resulting human being. This ideal opportunity arises before there has been any opportunity to consult the recipient of the enhancement.

Concerns about the timing of genetic enhancement are given provocative expression by the German philosopher Jürgen Habermas. Habermas argues against parental genetic enhancement of children on the grounds that a genetic enhancer "makes himself the co-author of the life of another, he intrudes—from the interior ... into the other's consciousness of her own autonomy." He continues, "the programming intentions of parents who are ambitious and given to experimentation ... have the peculiar status of a one-sided and unchallengeable expectation."[10] He proposes that this distinguishes genetic from environmental enhancement. The purported asymmetry between genetic enhancer and genetically enhanced undermines the equality characteristic of liberal societies. Habermas portrays the future generations of a society practicing genetic enhancement as "defenseless objects of prior choices made by the planners of today" and claims that "The other side of the power of today is the future bondage of the living to the dead."[11]

Habermas has identified a danger from enhancement that arises whether the means of enhancement is environmental or genetic. For example, the earlier one begins a program of deliberate enhancement the sooner one becomes world-class in that area. Fifteen-year-old chess players who have accumulated 10,000 hours of deliberate practice are likely to beat eighteen-year-olds with a mere 5,000 hours. Those who wait for children to achieve the capacity to make autonomous choices about the direction their lives will take will almost certainly lose out to parents whose educational programs have commenced earlier.

The Polgár girls speak of their enjoyment of chess. Even so, it's not as if they had much input into their educational program. László Polgár chose chess because it was the best way to demonstrate the efficacy of directed practice. He considered and rejected art and writing on the grounds that achievements in these areas tend to be more open to question. What some critics call a painting of genius others pronounce an ugly mess. If you checkmate your opponent in a game of chess, you're a winner, regardless of the opinions of any spectator. Parents less intent on proving the efficacy of extended deliberate practice are likely to make different choices for their children. But those choices may be equally indifferent to any desires of the children themselves.

Even very ambitious programs of directed practice may seem to give their human objects an option of resistance unavailable to the prenatally genetically enhanced. Had the Polgár girls insisted on hurling any proffered chess pieces across the room, then László and Klara might have abandoned their plan to turn their daughters into world-class chess players. Yet it's too late for a child to make it the case that her genome was never genetically altered.

This response overlooks an option of resistance available to the prenatally enhanced. While you in early adulthood may be unable to make it the case that the DNA of your embryo was never altered, you can prevent the alterations from having the effects your enhancer was hoping for. This opportunity is a consequence of the interactionist view of development according to which significant traits emerge not from the action of genes alone, but from the interactions of genes and environment. You can refuse to place the modified gene or genes in the environment necessary for them to have their intended effect. Suppose that you learn that your genome was altered with the intention of turning you into a brilliant mathematician. You're unlikely to become one if you refuse to study mathematics beyond grade school level. The mere act of genetically enhancing mathematical aptitude in no way prevents this choice.

(6) Are genetic enhancements more morally problematic than environmental enhancements because they are riskier?

The techniques that László and Klara Polgár used to enhance their daughters are novel. But these educational innovations should be acknowledged as significantly less novel than genetic modifications. Intensive homeschooling of the type described by Ericsson may lead to resentment, but it's unlikely to cause sudden death. When directed at an early embryo, genetic enhancement intervenes in processes that are foundational in human development. Our very preliminary understanding of how genes influence development should make us cautious about any alterations.

Genetic enhancement is riskier than variations of the environmental enhancement practiced by humans for millennia. This may not be true in the near future if we achieve a better understanding of the technologies of genetic modification. There are proposals to supplement human brains with electronic neuroprostheses and to inject self-replicating nanobots into human bloodstreams. When these enhancement technologies become available, environmental enhancements, considered collectively, could be more dangerous than genetic enhancements.

This section does not pretend to be exhaustive. My prediction, based on the parallel contributions of genes and environment to the construction of human beings, is that any purported moral differences will disappear when subjected to closer examination. We should be morally consistent in respect of enhancements. This consistency is one of *moral evaluation* and not of *moral conclusion*. We're not obligated to arrive at the same moral conclusion about a list of proposed genetic and environmental enhancements any more than someone selecting a sports team is required to choose equal numbers of black and white players. At a given time in the history of a given society, a specific collection of enhancements, genetic and environmental, will be available to its members. It's entirely possible the genetic enhancements available in the industrialized world of the early twenty-first century conflict more strongly with autonomy, open up more significant social divisions, or are associated with greater risks than available environmental enhancements. But this doesn't show that the epithets "genetic" or "environmental" pick out morally relevant properties. To return to the sporting analogy—a team that includes mainly black players may draw allegations of racially biased selection practices, but this pattern could arise simply because the particular black players who present themselves happen to be superior to the white ones. The pattern should reverse if next year's white candidates are better. Early twenty-first-century commentators make much of the potential for genetic technologies to radically recast human beings. If Kurzweil's predictions about the future of technology are accurate, then the attention of enhancers will soon turn to electronic enhancements presaged by current work in artificial intelligence. At this time, environmental enhancements should be the chief focus of moral investigation. They'll be the chief category of changes interfering with autonomous choice or fracturing the human species into haves and have-nots.

The Ideal of *Truly* Human Enhancement

Chapters 3 through 6 raised a variety of prudential and moral objections against radical enhancement. This chapter has been an indirect endorsement of certain forms of moderate enhancement. Some forms of environmental enhancement are clearly compatible with and conducive to human flourishing. It is inconsistent to refuse to extend that endorsement to some forms of genetic enhancement.

It can be prudent to undergo a process of moderate enhancement and morally good to arrange similar possibilities for your dependents. What emerges from the combination of this restriction and endorsement of enhancement is an ideal of truly human enhancement. The term "truly human" suggests a condition that nontherapeutic enhancements must satisfy. My use of the modifier "human" differs from its standard use in the enhancement debate. Human enhancement most straightforwardly indicates the enhancement of humans. As such, it differs from enhancement directed at members of some other species. For example, it would differ from murine enhancement—the enhancement of mice. I am interested in the manner of enhancement rather than the species of organisms that are its targets. To indicate this different interest I use the term *truly human* enhancement. Given that I'm interested in truly human enhancement of humans, a more accurate but somewhat stylistically clumsy label might be truly human, human enhancement. The ideal of truly human enhancement rejects many enhancements as not prudentially valuable to humans—they are inhuman, human enhancements.

Some differences between human and inhuman enhancements are differences in kind. Inhuman types of enhancement include those that enhance our underwater performance by fitting us with gills or that enhance our tree-climbing abilities by the addition of prehensile tails. Other inhuman enhancements differ in degree. Many human enhancements become inhuman when they exceed a certain degree. To say that they become inhuman is to say that we are entitled to value them less than "more human" moderate enhancements.

Are there prudential differences between enhancements that are inhuman because of a difference in kind and enhancements that are inhuman because of a difference in degree? I suspect that there is a danger associated with enhancements that differ in degree that is largely absent from enhancements that are different in kind. Take, for example, a modification that grafted wings to the human body, granting humans the power of flight. The fact that this enhancement does not replace an existing, less developed ability removes one threat to our identities. The enhancement will not make former achievements seem less valuable by subjecting them to an evaluative standard that makes them seem impressive. The ability to fly by flapping our wings brings new kinds of experience that need not displace existing experiences. The point about the expected instrumental value of such a procedure remains. Those who would enable humans to fly by grafting wings to our torsos face immense challenges in working out how to integrate wings into human bodies. An approach likely to yield more substantial instrumental benefits is to try to make jetpacks safe. These should carry us into the skies in a way that offers many of the thrills of wings.

The ideal of truly human enhancement conforms to the anthropocentric ideal. Enhancements have prudential value relative to human standards. Some enhancements of greater objective magnitude are less valuable than enhancements of lesser magnitude. Moderate enhancements do not exceed our powers of imaginative engagement. This type of enhancement has the potential to bring us into closer contact with distinctively human internal goods. It does not erect imaginative barriers. It does not make our former achievements seem worthless.

8   Why Radical Cognitive Enhancement Will (Probably) Enhance Moral Status

The chief focus of the book's discussion so far has been on the individuals who have undergone radical enhancement. Radical enhancement is prudentially irrational—it is predictably bad for those who undergo it. Lesser degrees of enhancement—moderate enhancement—can promote the interests of those who undergo them. It can be prudentially rational to submit to moderate enhancement. This chapter switches focus to the domain of morality. Do we have reason to morally condemn too much enhancement?

This chapter and the one that follows it advance two claims about human enhancement. Some degree of biotechnological or cybernetic enhancement could enhance the moral status of human persons. It is likely to lead to post-persons—beings with moral status higher than persons. On the way to this conclusion, I respond to a challenge to the possibility of post-persons from Allen Buchanan.[1] Chapter 9 proceeds from the possibility of post-persons. It presents a case for the moral wrongness of enhancing moral status. We should look upon moral status enhancement as creating especially needy beings. The predictable consequence of creating beings with a moral status higher than personhood is that deserving needs of human persons will go unmet. We are not required to create morally enhanced beings and so we should not.

Why does Buchanan dispute the possible existence of post-persons? He allows that genetic and cybernetic technologies may succeed in enhancing the powers relevant to personhood. Buchanan observes that we find it difficult to describe statuses higher than personhood. This apparent inexpressibility does not bedevil attempts to describe other human enhancements. It's relatively clear what kinds of change would have the effect of enhancing mathematical expertise to levels beyond those currently achievable by human mathematicians. Someone with enhanced status as a mathematician could perform proofs impossible for humans. They could perform in their heads calculations that humans could not perform without assistance from a computer. Higher moral statuses differ from higher statuses as mathematicians by being difficult to describe or imagine. A natural explanation for this inexpressibility is that there are no such statuses and nor could there be.

This chapter offers an explanation of the apparent inexpressibility of post-personhood that is compatible with the possible existence of post-persons. Moral statuses higher than personhood can be viewed as analogous to objects including space-time singularities whose existence we can infer without our being able to directly observe them. The apparent inexpressibility of post-personhood says something about us, and not about the possible existence of a moral status superior to personhood. It is an implication of accounts that make a cognitive capacity, or collection of such capacities, constitutive of moral status, that those who do not satisfy the criteria for a given status find these criteria impossible to adequately describe. I offer an inductive argument that compensates for our limited powers of moral imagination. The possible existence of post-persons can be inferred from our observations about the different moral statuses of persons, sentient nonpersons, and nonsentient things.

Enhancing Moral Status versus Enhancing Moral Dispositions

There are two ways in which the moral enhancement of human beings might be sought. Humans could undergo either *moral disposition* enhancement or *moral status* enhancement. For brevity's sake, I shall sometimes refer to the first as disposition enhancement and the second as status enhancement.

The aim of moral disposition enhancements is to increase the moral value of an agent's actions or character. A variety of proposals are in the offing. Mark Walker offers an account of moral disposition enhancement that focuses on character traits. He proposes a program to make citizens more virtuous.[2] Julian Savulescu and Ingmar Persson describe how we might enhance moral dispositions by boosting empathy and cooperativeness.[3] Thomas Douglas argues that we increase this moral value by attenuating certain countermoral emotions.[4] In this book, I advance no claims about the feasibility or justice of disposition enhancements.

The aim of moral status enhancement is not to increase the moral value of our actions or characters. Rather, it increases a being's entitlement to certain forms of beneficial treatment and reduces its eligibility for certain forms of harmful treatment. The following elaboration of this basic idea is due to Buchanan. He proposes that a being has moral standing "if it counts morally, in its own right."[5] Moral status differs from moral standing in being a comparative notion. Suppose that two beings both have moral standing. One may have higher moral status than the other. That is, one being may count for *more* morally in its own right than the other.

Much of the discussion in this chapter concerns a particular moral status—that indicated by the term "person." The concept of personhood that occupies the central location in Buchanan's discussion is a Kantian one, according to which a person is a being with a capacity for practical rationality.[6] Persons can both be held accountable and hold others accountable. While much of my discussion assumes this account of personhood, the points I make should apply to a Lockean account that identifies persons as rational, self-conscious beings, who are aware that they have interests that persist over time. They apply to any account that makes a cognitive capacity essential to personhood.

One might enhance the moral status of a sentient nonperson by introducing into it the cognitive traits sufficient for personhood. For example, suppose one were to make cybernetic modifications to a sheep, giving it mental powers identical to those of human persons. It's possible that the precise modifications of cognitive and affective powers required to turn sheep into Kantian persons will differ from those required to turn them into Lockean persons. Each view should allow, depending on the precise nature of the psychological enhancements, that the sheep has undergone status enhancement. It should now qualify for the moral protections due to persons.

Why It's So Difficult to Enhance the Moral Status of Persons

It's possible to enhance status up to the level of personhood. But is it possible to enhance status beyond this point? Buchanan's serious doubts about the possibility of enhancing the status of human persons derive from a claim he calls the moral equality assumption. This is the idea that "all who have the characteristics that are sufficient for being a person have the same moral status."[7]

I follow Buchanan in using the term "mere person" to indicate a being who satisfies the criteria for personhood but fails to satisfy any criteria for a higher moral status. Post-persons would, on this understanding, be persons, but not *mere* persons. So why does Buchanan find it unlikely that such beings could exist?

Buchanan presents personhood as a *threshold concept*. Giving a being who does not meet the criteria for personhood greater powers of practical rationality may improve its moral status by enabling it to satisfy those criteria. Enhancements beyond this point make no difference to whether or not a being satisfies these criteria. They should therefore not place it in a moral category superior to that of persons. Buchanan says, "If a person's capacity for practical rationality or for engaging in practices of mutual accountability or for conceiving of herself as an agent with interests persisting over time were increased, the result presumably would be an enhanced person, not a new kind of being with a higher moral status than that of person."[8]

A Justification for (Talking about) Moral Statuses

The arguments of this and the following chapter assume the existence of moral statuses. I follow Buchanan in understanding moral statuses as thresholds. This is actually somewhat controversial. Some philosophers dispute appeals to moral statuses and their implied thresholds in philosophical arguments attacking or defending enhancement.[9] I offer the following pragmatic justification for conducting my argument purely in terms of moral statuses. I am a defender of the idea that enhancement could boost a human's entitlement to benefits and protection against harms. The assumption of moral statuses is difficult to square with this possibility.

Alternatives to moral statuses seem friendlier to moral enhancement. For example, Allen Buchanan considers an interest-based account that ranks beings according to the differing amounts or degrees of good their lives involve. If interests are to take the places of statuses, then enhancement needn't produce a moral *status* superior to personhood. But it could produce beings whose weightier interests give them superior moral protections against harm and entitlements to benefits. We might, for example, decide to sacrifice a dog before a human not on the grounds that it possesses an inferior moral status but instead because humans' superior rationality and imaginative abilities give them a stronger interest in surviving. The dog's more limited cognitive and imaginative capacities mean that it is less harmed by death than is the human. Presumably the being with powers of reason and imagination far superior to those of humans anticipates death in a more vivid way than we do and therefore has a stronger interest in avoiding death. Harms that fall short of death derail post-persons' projects that are more complex than any of those that we undertake.

So, in granting the existence of moral statuses, I grant an assumption that is hostile to the existence of beings with stronger entitlements to benefits or protections against harms. It's for Buchanan to defend moral statuses and their associated thresholds—I can concede their existence.

Three Obstacles to Moral Enhancement

There are three obstacles to acknowledging a moral status higher than personhood.

(1) The problem of the logic of thresholds: When used to indicate moral status, personhood is, according to most analyses, a threshold concept. Once one satisfies it, additional increments of the properties relevant to satisfying it make no moral difference. How can a higher moral category exist if personhood corresponds with a threshold?

(2) The problem of how to improve upon inviolability: According to an analysis favored by Buchanan, persons differ from nonpersons in being morally inviolable. Suppose that we accept this analysis. How could enhancement improve the moral treatment of a being whose most fundamental rights already cannot be violated?

(3) The problem of expressing moral statuses higher than personhood: What are the criteria for moral statuses higher than personhood? It's difficult to imagine what the criteria for post-personhood might be. Attempts to specify them seem to succeed only in adding to the powers of persons in ways that enhance them but make no difference to their moral status. It's relatively easy to imagine enhancements that make persons more intelligent. What's difficult is seeing how these changes could enhance moral status.

(1) The Problem of the Logic of Thresholds

Buchanan presents the concept of moral status indicated by personhood as a threshold concept, not a scalar concept. Wealth is a scalar concept. The degree of one's wealth increases with the acquisition of additional quantities of money and items of monetary value. There's no point at which additional quantities of money or valuable items cease to make a difference to one's wealth—one can always become wealthier by acquiring more money or items of value. Having the moral status of a person is, in contrast, a threshold concept. Buchanan says that "according to theories that accord moral status (or the highest moral status) to persons, understood as beings who have the capacity for practical rationality or for engaging in practices of mutual accountability, what matters is whether one has the capacity in question. Once the threshold is reached, *how well* one reasons practically or *how well* one engages in practices of mutual accountability does not affect one's moral status."[10]

The threshold view has much appeal. It explains an observed moral equality among persons who satisfy the criteria for personhood. Among those who satisfy the criteria, there's wide variation in the relevant capacities. It is apparent that some human persons are better than others at reasoning practically. Yet we strongly resist acknowledging many moral statuses to correspond with different levels of attainment in practical reasoning.

This chapter's discussion assumes that moral statuses exist and that some statuses, including personhood, correspond to thresholds. Buchanan argues that the recognition of personhood as a moral status rules out the possibility of enhanced moral protections against harm and entitlements to benefit. I respond that this assumption is compatible with moral enhancement.

In what follows, I explore two ways in which moral statuses might explain the observed equality of all human persons. One posits what I will call a *strong threshold* in moral status. The other posits *weak thresholds*. Weak thresholds are compatible with the enhancement of moral status beyond personhood.[11]

A strong moral status threshold A point or region beyond which *no* improvement to the capacities relevant to moral status makes any difference to status.

A weak moral status threshold A point or region beyond which *moderate* improvements to capacities relevant to moral status make no difference to status. Improvements of greater magnitude *could* make a difference to status.

Buchanan's moral equality assumption posits a strong threshold in moral status. A weak threshold differs in allowing that improvements to the capacities relevant to moral status could make a difference to status so long as these improvements are more than moderate. If we understand moderate improvements as encompassing the entire observed human range above the minimum criteria for personhood, then the existence of a weak threshold in moral status is compatible with the observed moral equality of all human persons. Positing a weak threshold in moral status permits but does not settle the question of whether enhancement of practical reason well beyond the human range might have this effect.

The notion of a weak threshold may seem less intuitive than that of a strong threshold. It's not as theoretically tidy. But it's a notion that we make frequent use of. We often place more than one threshold along a single axis of human ability. Since the placement of one strong threshold precludes the placement of a second strong threshold, these must be weak thresholds.

Consider an example from outside the moral domain. Suppose you decide to improve your Spanish language abilities. You sign up for classes at a school which administers a test. The test uses weak thresholds to place you in an appropriate class. If you know no or close to no Spanish, you go into an introductory class. More knowledgeable students are placed in an intermediate class. Students with the best language skills enter an advanced class. There is some variation in the language abilities of students who find themselves in the intermediate class. Some barely avoid placement in the introductory class; others fall just short of the standard required for enrolment in the advanced class. The existence of a weak threshold means that variation in language abilities over this range makes no difference to the class in which a student is placed. But variation of greater magnitude does.

If we suppose that personhood indicates a weak threshold in moral status, then the logic of thresholds does not preclude the existence of moral statuses higher than personhood.

(2) The Problem of How to Improve upon Inviolability

If persons are already morally inviolable, then what higher form of respect is due to post-persons? How could the respect we owe to morally inviolable persons differ from what we owe to post-persons?

As part of his argument that post-persons (McMahan's preferred term is "supra-persons") are possible, Jeff McMahan proposes that we reject an absolutist reading of inviolability according to which there are no circumstances in which it could be right to sacrifice an inviolable being.[12] Instead, we should allow that violability and inviolability come in degrees. It's conceptually possible to create beings who are more inviolable (or less violable) than persons.

McMahan argues that this approach explains commonsense verdicts about what can and cannot be done to morally inviolable beings. It is almost never permissible to sacrifice a human person to produce some other good. But there is no absolute prohibition on sacrificing persons. In conditions of extremity—McMahan's example involves killing the innocent as the only way to prevent the killing of a very large number of other innocents—it is permissible to sacrifice a person. This and other intuitive judgments about inviolability are best explained by attributing to humans a very low degree of violability (alternatively, a very high degree of inviolability) rather than ascribing absolute inviolability. This analysis of violability would permit post-persons to differ morally from persons by having an even lower degree of violability.

McMahan gives an example that illustrates this higher degree of inviolability. He supposes that there is some number of innocent lives for which it would be right to sacrifice an innocent person. McMahan reasons that the lower violability of post-persons could make it impermissible to sacrifice them for the purpose of saving this number of innocent humans.

(3)   The Problem of Expressing Moral Statuses Higher Than Personhood

We come now to the most serious obstacle to higher categories of moral status. There seems to be a significant barrier in grasping the criteria that one must satisfy to be correctly pronounced a post-person. It's easy to imagine beings who are more intelligent than we are. But it's difficult to see how this greater intelligence could place them in a higher moral category.

Buchanan allows that this point is not decisive. He urges that we not "confuse a failure of imagination with conceptual incoherence."[13] But he nevertheless says that "in the absence of an account of what the higher threshold would be like, the claim that there could be beings at a higher threshold who would have a higher moral status is not convincing."[14] Those who assert the possibility of higher moral statuses owe either a description of the criteria for a higher moral status or an explanation of why the great difficulty in producing such an account might nevertheless be compatible with the existence of such categories. In what follows, I offer an explanation for the apparent inexpressibility of moral statuses higher than personhood. I argue that, if these criteria are constituted by a cognitive capacity or collection of cognitive capacities, then it's reasonable to expect that they will be difficult for mere persons to formulate. This fact notwithstanding, it is possible to infer their probable existence. It should be viewed as improbable that there would be no categories of moral status higher than personhood.

Three Attempts to Describe Higher Moral Statuses

In this section, I explore three attempts by philosophers to overcome what I will call the expressibility problem—the problem of describing moral statuses superior to personhood. Each of these philosophers offers what I will call a constructive account of moral status enhancement. By constructive, I mean that they seek to actually describe moral statuses higher than personhood. These authors offer accounts that purport to make clear how and why moral status enhancement produces beings with a superior entitlement to certain benefits and a reduced eligibility for certain harms. In each case, they make an attempt to say how persons might be improved in ways that endow them with higher moral status. I will argue that each explanation fails to overcome the expressibility problem.

Here is a general feature of the following constructive accounts of postpersonhood: they tend to link the variety of moral enhancement we have identified as moral status enhancement with the variety of enhancement called moral disposition enhancement. The shared label "moral enhancement" suggests such a link. The shared label is unfortunate. I argue against a strong connection between the aim of moral disposition enhancement, which is to increase the moral value of an agent's actions or character, and moral status enhancement, which increases a being's entitlement to certain forms of beneficial treatment and reduces its eligibility for certain forms of harmful treatment. Those who perceive a link between the two forms of enhancement tend to minimize the difficulties in imagining status enhancement. This is because the task of imagining moral disposition enhancement is rather straightforward—especially for those who have a preferred account of morality. One simply imagines changes to human beings that enhance the moral value of their actions or characters. Such assessments would be guided by a preferred account of morality. This is, of course, not to downplay the difficulty in discovering interventions that would bring about these changes. Nevertheless, it seems to face few of the imaginative obstacles that beset attempts at imagining moral status enhancement. All but one of the following constructive accounts of moral status enhancement make disposition enhancement a significant constituent. They thereby underestimate the difficulty of describing higher moral statuses.

The mistake of linking moral status enhancement with moral disposition enhancement has a probable source. It is seemingly suggested by the Kantian view about personhood. For Kantians, personhood is both a status and a disposition. To identify a being as a person is to say something about how it should be treated. Persons are ends in themselves; they cannot be treated purely as means. Personhood is also a moral disposition. Persons can accept the guidance of moral reasons. This linkage of a moral status and a moral disposition implies nothing about the effects of enhancement. The enhancement of a moral disposition need not enhance status. Compare: There is a connection between a bird's means of flight and its means of thermoregulation. Its feathers both help it to fly and help it to keep warm. But enhancing the capacity of feathers to generate warmth does not necessarily enhance their propensity to promote flight. Indeed, it's likely to have the opposite effect.

The existence of a strong link between moral disposition enhancement and moral status enhancement would be bad news for current advocates of disposition enhancement. Suppose moral disposition enhancement could eradicate some portion of human selfishness. We might expect dispositionally enhanced humans to be better at placing the more considerable interests of others ahead of their own. However, suppose also that the subjects of disposition enhancement also find themselves acquiring an enhanced moral status. The attempt to improve treatment of others could backfire. It could replace immoral harmful treatment of others with harmful treatment that is justified because of a difference in moral status. Those who formerly violated the rules of morality in placing their own less significant interests ahead of the more significant interests of others will find that their enhanced status has boosted the significance of their interests. The suffering that they inflict would be justified rather than unjustified. This should be a concern for advocates of moral disposition enhancement. Those who worry about the high level of unjustified suffering in the world would like this suffering to be eliminated or at least reduced. They are typically not attracted by the possibility of changing the world in ways that make little difference to the quantity of global suffering, but tend to convert unjustified suffering into justified suffering.

DeGrazia's Dispositionally Superior Post-Persons

David DeGrazia offers for consideration a scenario in which humans might have an enhanced moral status.[15] I quote his scenario in full.

A Future with Post-persons. It is 2145. Out of massive human population, a discrete population has evolved, through carefully planned genetic modifications, and has achieved a considerable number. These beings are in many respects superior to unenhanced people. They typically learn 10–12 human languages, a feat made possible by their retention throughout their lifetimes of the sponge-like capacity that young human children have always had. Their memories, on average, are as capacious as those considered prodigious among the unenhanced population. They have far more extensive self-awareness than ordinary persons, being able to detect with little or no effort the ways in which their biological endowment, early environment (which they remember very clearly) and present environment create myriad dispositions and pressures to think and behave in particular ways. Being far more rational than ordinary people, they are embarrassed to have evolved from a type of creature so susceptible to superstitions, myths, cultural prejudices, ethnic and religious discrimination, unconscious bias in favour of one's own interests, a litany of logical fallacies and so on. They marvel at the way even the philosophers and scientists among the unenhanced population regularly deceive themselves about their own strengths and weaknesses, their motives and the likelihood of adhering to resolutions. Bringing together several of these strengths, the post-persons are vastly superior in their moral capacities. First, they are consistently impartial whenever impartiality is morally required. Second, because they screen out distracting stimuli and think very quickly, they reach correct moral judgements in conditions of stress no less consistently than they do in leisurely reflection. Third, they suffer from weakness of will so seldom that any of their members who does so is regarded as having a psychological disorder. Finally, in comparison with ordinary persons, these enhanced humans are enormously adept at envisaging the likely consequences of their choices and identifying the implications of their moral judgements.[16]

The members of this subpopulation are superior to unenhanced humans in a variety of ways. DeGrazia makes a good case for their having superior moral dispositions. They avoid many of the moral mistakes of their unenhanced cousins. Does this support DeGrazia's further claim that they have an enhanced moral status?

DeGrazia offers some considerations in support of a higher moral status. He says that his candidate post-persons are superior to contemporary

human beings "in ways that matter to us."[17] Moreover, they "regard themselves as *different in kind* from persons."[18] The central actors in DeGrazia's
story are much more intelligent than us. But it's far from clear that this difference must be reflected in a difference in moral status. The concept of personhood encompasses cognitive abilities beneath those of normal humans.
Mildly mentally handicapped humans are inferior "in ways that matter to
us" without possessing a lesser moral status. It's unclear why the category of
personhood shouldn't extend beyond human cognitive norms to encompass DeGrazia's genetically enhanced beings. Their superior philosophical
acumen may mean that mere persons should accept their judgments were
they to sincerely credit themselves with a higher moral status. But the truth
of this conditional claim does nothing to support the proposition that they
would actually judge themselves to have a higher status. We do not mark the
difference in intelligence between cognitively normal humans and mildly
mentally disabled humans with a difference in moral status.

When I originally made this point, DeGrazia responded to my claims
about the judgments of cognitively and dispositionally enhanced beings.[19]
He pointed out that his thought experiment did not involve "a counterfactual
claim as to what such beings would actually think"; rather, he built "into the
thought-experiment the stipulation that they would have this perception."
Furthermore, "this perception would not be unreasonable, based on the differences between post-persons and persons."[20] I am suspicious of DeGrazia's
claims about what he's permitted to stipulate in a thought experiment of this
type. It's true that one can stipulate many things in a philosophical thought
experiment. Philosophical thought experiments involve many improbable
scenarios. But one thing that should not be stipulated is the conclusion that
a thought experiment is designed to establish. Judith Jarvis Thomson can
stipulate all of the descriptive details of her famous thought experiment in
which a passerby awakes to find herself hooked up to a medically needy
violinist.[21] But one thing that she cannot stipulate is that it is morally right
that the passerby unhook herself. This is the philosophical conclusion that
the thought experiment purports to establish. It might be understandable
that DeGrazia's putative post-persons believe themselves to have a higher
moral status, but it will be mistaken for them to do so if the many differences between the enhanced and unenhanced do not correspond with a
difference in status. The members of slave-holding classes certainly acted as
if they had a moral status higher than that of their slaves. Moreover, it could

be reasonable for someone born into that class to believe that he had such a higher status. But these facts do not establish an actual difference in status. They correspond with no actual difference in moral status.

What of the putative connection between moral disposition enhancement and moral status enhancement? I'm suspicious of any strong connection. For example, I believe that consequentialism offers the best account of normative ethics. I freely concede, however, that I'm a fairly haphazard consequentialist. In many situations I choose actions whose predictable consequences are worse than other actions I might have performed. I know of more reliable consequentialists. For example, there's Zell Kravinsky, the man whose consequentialism has led him to donate a kidney to a stranger and who stands prepared to donate a second kidney should morality direct him to do so.[22] And then there's the animal welfarist philosopher Peter Singer, far better than me at consistently excluding from his reasoning morally arbitrary facts about species membership. Unlike Singer, I do not give 25 percent of my salary to charity. Kravinsky and Singer may be more reliable and therefore dispositionally superior consequentialists. What's difficult to see is how this should take a step in the direction of endowing them with a moral status superior to personhood.

For further evidence of the lack of a connection between the two varieties of moral enhancement, consider another possible subpopulation in DeGrazia's world of 2145. This subpopulation is the handiwork of sadistic genetic engineers bent on moral disposition pejoration. They design beings more reliably immoral than any member of the unenhanced population. Unlike haphazardly immoral unenhanced humans who tend to specialize in certain forms of immorality, these genetically engineered beings have a thoroughgoing commitment to evil. They consistently choose actions with the worst consequences. It's easy to see why the people of 2145 would regret the existence of this subpopulation. But there's no reason to think that the moral disposition pejorations must have an effect on personhood. They need not turn their subjects into pre- or subpersons, beings with a moral status inferior to persons.

McMahan's Freer, More Conscious Post-Persons

In a paper primarily concerned with exposing inconsistencies in our views about beings who fall short of the criteria for personhood, Jeff McMahan offers two tentative descriptions of beings who might possess moral statuses higher than mere persons. On the way to these accounts he offers a helpful suggestion about the source of the barrier to understanding that plagues attempts to describe higher moral statuses.[23] The properties that account for supra-persons (McMahan's term for post-persons) may be emergent properties. Emergent properties are typically higher-level properties that present as a result of combinations of lower-level properties. They do not result from the simple augmentation of lower-level properties. One example of emergence is the sweet taste of glucose. This sweetness emerges from the specific combination of carbon, oxygen, and hydrogen atoms that make up glucose molecules. The carbon, oxygen, and hydrogen atoms are not themselves in any way sweet. The sweetness of glucose does not result from combining elements that are themselves somewhat sweet to get a chemical compound that is very sweet. Rather, the emergent property of sweetness comes into existence with the specific combination of these elements. If supra-personhood is a property that emerges from specific combinations of enhancements of human capacities, then we humans might be powerless to predict it. The fact that these are emergent properties may explain failures of imagination. McMahan makes the point that "the psychological capacities that we have that are reasonable candidates for the basis of our higher inviolability—self-consciousness, the ability to act on the basis of reasons, and so on—seem to be emergent properties that have arisen from the combined enhancement of capacities found in animals."[24]

McMahan does go on to offer a couple of descriptions of post-persons. His first suggestion involves beings who might be freer in some deeper sense than we are. McMahan notes that one of the differences between persons and animals that is frequently cited as morally significant is that persons possess free will and animals do not. According to the libertarian account of free will, our free choices are exceptions to the deterministic laws that govern events outside of our wills. This purported fact about us is sometimes cited as an explanation for our high moral status. Libertarians' philosophical opponents either deny that we are free or insist that our freedom is compatible with determinism. McMahan continues:

But suppose that the notion of libertarian free will is actually coherent. And suppose further that those who have asserted that libertarian free will is what distinguishes us morally from animals are right that it is a sufficiently significant capacity to ground a difference in moral status, but wrong to believe that we actually possess it. But suppose, finally, that supra-persons would have it. They would have a psychological capacity that we lack but that most people have believed that we have and that is what distinguishes us morally from animals.[25]

If libertarian free will would suffice to give persons a moral status higher than beings who lack it, then, supposing that we lack it, it should suffice to give beings who actually do possess it a status higher than us.

McMahan is aware that this proposal is contingent on the philosophical coherence of libertarian accounts of free will. This constructive account of moral status enhancement cannot be more coherent than the account of freedom that it assumes. For example, the "higher" freedom of beings with libertarian free will depends on an adequate response to the conceptual objection that agents whose choices were literally uncaused would not be freer than us. Rather, they would be less free. Their choices would be no more responsive to their wills than would be the activity of some of the subatomic particles in their brains.

Suppose that libertarian free will is philosophically coherent, that beings with libertarian free will are realizable in our universe, and that human mere persons do possess this variety of free will. There remains the further, separate issue as to whether mere persons, beings free in the restricted sense as compatible with determinism, could use enhancement technologies to turn themselves into beings with libertarian free will. Consider the most florid vision of the cognitive enhancements brought by enhancement technologies—the fabulously powerful minds that Kurzweil presents as consequences of the exponential improvement of our cognitive powers. It's difficult to see how the addition of new computational powers or the augmentation of existing computational powers might abruptly generate this new phenomenon of libertarian free will, somehow fashioning it out of the humdrum variety that is compatible with determinism. Perhaps it might happen at the point in our cognitive futures when our minds become quantum computers and they acquire the ability to fully exploit the computational potential of the subatomic realm?

In any event, one may agree with McMahan that nothing rules out the existence of beings with libertarian free will and that such beings would be post-persons without believing that any application of an enhancement technology could transform us from beings with compatibilist free will into beings with libertarian free will. This form of moral status enhancement need not preoccupy those who hope to radically enhance human cognitive abilities.

McMahan has a second suggestion that may avoid some of the meta-physical obstacles of his first suggestion. His second class of supra-persons seems accessible to enhancement technologies. They seem like beings that human mere persons could become:

> It is widely held that empathy is relevant to, and perhaps even necessary for, moral agency, and many philosophers have held that the capacity for moral agency is necessary for the higher form of moral status. Suppose, then, that supra-persons would have a capacity that would be better for moral agency than mere empathy. Suppose they could actually experience other individuals' mental states while simultaneously reflecting on those experiences in a self-conscious manner from their own point of view. This would require a divided form of consciousness, but that would be only a rather extreme instance of the fragmentation of consciousness of which we are increasingly aware in ourselves.[26]

Do McMahan's admittedly preliminary comments answer the problem of describing higher moral statuses? It seems clear that this expanded consciousness has the potential to enhance moral dispositions. It may not always have this effect. Sadists might derive enjoyment from a capacity to occasionally inhabit the consciousness of their victims. But it does seem that an inability to fully understand the consequences of our actions for others is a cause of much harm. People would be less likely to be indifferent to the suffering of people in distant lands if they could experience their suffering. Indifference about the effects of poverty would be less likely if we could directly sample some of its psychological effects. What is less clear is how this would bring about moral status enhancement.

There are some natural experiments in the sharing of consciousness with other beings. Consider Krista and Tatiana Hogan.[27] Krista and Tatiana are craniopagus conjoined twins. They are joined at the head—a neural bridge links Krista's thalamus with the thalamus of Tatiana. The twins are different individuals—Krista and Tatiana do not share a single mind. But they seem capable of conscious awareness of some of each other's sensations. Tatiana seems aware of some of the sensations that accompany Krista's eating and drinking. Tatiana might have a self-interested reason for not inflicting pain on Krista that does not govern the interactions of other siblings.

The question of interest here concerns whether the Hogan twins constitute some step on the way to post-personhood. The Hogans manifest a more restricted expansion of consciousness than that described by McMahan. The twins are limited to each other's conscious awareness. They cannot directly sample the consciousness of any other sentient being with whom they might interact. Perhaps study of the twins could enable the invention of transferable neural bridges capable of connecting the thalamus of its possessor with someone else's thalamus. Such a modification might dramatically enhance its subjects' moral dispositions. But it's less clear why it would enhance moral status.

Douglas's Enhanced Cooperators

Thomas Douglas offers a further scenario in which enhanced humans acquire a superior moral status. He suggests that a moral status higher than personhood might "be conferred by the capacity for constructive participation in some new form of social co-operation."[28] Douglas continues: "It is not inconceivable that quantitative increases in capacities for altruism, self-control and general intelligence might lead enhanced beings to develop new and qualitatively different forms of social co-operation."[29] Douglas supposes that our political and legal institutions might be replaced by new institutions available to beings with greater altruism and enhanced abilities to predict and plan cooperative interactions. He extrapolates from nonhuman cases: "After all, the inability of non-human animals to constructively participate in characteristically human forms of co-operation (democracy, economic markets, the operation of legal systems and so on) is one of the more plausible (or less implausible) grounds for assigning those animals sub-personal moral status."

I do not want to seem too much like a broken record here. I agree that Douglas has described a path to dispositional enhancement. We would collectively benefit from the new forms of social interaction that he is describing. What is less clear is why these new forms of cooperation would enhance moral status. For example, we may be witnessing the beginnings of these superior patterns of relationship in Facebook and Google+. These innovations seem to have enabled modes of cooperation unavailable to those with limited social contacts outside of their traditional communities. Praiseworthy though they may be, they do not seem to be enhancing moral status. Science fiction presents for consideration many human societies that have replaced economic relationships with morally superior relationships. The human societies of *Star Trek* have supposedly done away with money. It doesn't seem, when they indulge in their frequent time travels back to our era, that we witness interactions between two different moral statuses of human beings. It doesn't seem that the crew of the time-traveling starship *Enterprise* would be morally justified in preferentially sacrificing the lives of twentieth- and twenty-first-century humans to save the lives of their shipmates.

I should be clear. The objection is not that none of the DeGrazia, McMahan, or Douglas scenarios actually describes moral statuses higher than personhood. They may do. I have argued that we—as human mere persons—cannot see how or why the beings they describe could have a moral status higher than our own. Perhaps the craniopagus twins Krista and Tatiana Hogan have inadvertently taken the first step toward moral status enhancement. Beings whose insight into the experiences of others is not limited to a sibling but instead is freely transferable from mind to mind may have turned the twins' modest beginnings into genuine moral status enhancement. The point is that it is difficult to see how facts about the twins or empathetically more powerful beings would constitute evidence for status enhancement that we should believe. To return to an example that featured in chapter 5, those who find "42" to be a disappointing answer to the "Ultimate Question of Life, the Universe and Everything" should not complain that it could not be true. Rather, they should complain that those to whom the answer is given have no way of connecting 42 to the concepts we use to address questions about the meaning of life and ultimate rationales for the existence of the universe. There are no obvious references to the existence, or nonexistence, of supernatural beings. There's no attempt to explain the whys and wherefores of the big bang or the existence of a life-sustaining universe. The request to find the question to which "42" might be a satisfactory answer is the request to find those conceptual and explanatory connections. The proposals of McMahan, DeGrazia, and Douglas are rather like Deep Thought's answer. They don't explain why the given enhancements and improvements enhance status. If my hypothesis about the inexpressibility of moral statuses higher than personhood is correct, then mere persons may be forever denied such a satisfactory explanation.

Criteria for Higher Moral Statuses and the Expressibility Problem

We seemingly have little difficulty in describing many of the things that cognitively enhanced beings might do. They might perform fiendishly complicated mathematical calculations in their heads; they might take minutes to read and internalize the entire contents of the twenty-volume edition of the *Oxford English Dictionary*, and so on. We can't do these things, but we have little difficultly in imagining much of what doing them would involve. Of course, some mental performances of enhanced humans might not occur to us. But if we were to be alerted to the possibility of these performances, then there should be no obstacle to imagining them. These imaginings may not be particularly detailed—but they should be sufficiently detailed for us to be reasonably confident that what we're attempting to imagine is, in fact, possible. There is, however, a species of enhancements which causes a systematic failure of our imaginative powers. We can't imagine enhanced humans doing logically impossible things—such as proving that the square root of 2 is a rational number. And this is something that no degree of cognitive enhancement could permit. If we can't imagine the improvement of cognitive powers enhancing moral status, then might it, like proving the rationality of the square root of 2, be impossible?

In what follows, I argue that we cannot express these criteria because they are constituted by capacities that are *cognitive*. The fact that criteria for post-personhood are cognitive is a barrier to mere persons' powers of expression and imagination. We, nevertheless, have reason to believe in them.

The Kantian analysis preferred by Buchanan places beings in the category of persons by virtue of a cognitive capacity—one is a person by virtue of the capacity for practical reason, the capacity to engage in practices of mutual accountability. It's a feature of a criterion determined by a cognitive capacity that those who do not satisfy it are typically unable to properly understand it. If we really understood the forms of practical reasoning that were constitutive of post-personhood then we'd be able to perform them and as such would satisfy the criteria for post-personhood. Even error-riddled, halting performances of these cognitive tasks should qualify us for inclusion in this moral category. Our fellow post-persons may view us as particularly dull members of their moral category, but they should, nonetheless, accept us as their moral equals.

The Kantian approach to moral status says nothing about how well one performs the cognitive tasks constitutive of personhood—for example, the facility with which one can universalize subjective principles of action. Teachers of Kantian ethics are aware that some human persons perform these tasks more easily than others. But even those who can reason practically only with much effort and with many mistakes satisfy the criteria for Kantian personhood. The same points should apply to cognitive tasks constitutive of post-personhood.

We should not underestimate the manner of understanding that the Kantian approach to moral status supposes. It's possible that a being who just fails to satisfy the requirements of higher moral status would notice some of the consequences of having a lower status. This being might notice that beings that resemble it are frequently caused to suffer by beings that differ from it. This increased probability of suffering could be a consequence of the being's possessing a lower status—higher-status beings might, correctly, be sparing themselves suffering by redirecting it to lower-status beings. The lower-status being might make such observations without truly understanding why it has a lower status. We might make such observations of the actions of post-persons while understanding very little about how they justify their choices. We would fail to possess a moral status implied by that understanding.

Consider an analogous case in respect of mathematical understanding. Someone asked whether she *really* understands Gödel's incompleteness theorem is being asked to do more than just confidently assert the proof shows that arithmetic is incomplete. She is being asked whether, if handed pencil and paper, she could perform the key steps in the proof. Among those who really understand Gödel's proof, some produce those steps with greater ease than others. But all who claim a genuine understanding ought to be able to produce them under some circumstances.

So it's at least compatible with our apparent cluelessness in respect of higher moral statuses constituted by a cognitive capacity that they do exist but we are unable to properly describe them. Of course, it's also compatible with our apparent cluelessness that there are no such criteria. In what follows, I present a moderately strong inductive argument for the existence of such criteria.

Why Cognitively Enhanced Beings Are Probably Better Than Us at Judging Relative Moral Status

How can we overcome the expressibility problem? I have argued that the fact that higher moral statuses are constituted by a cognitive capacity or a collection of such capacities means that we who do not satisfy these criteria cannot adequately describe them. We can, in principle, overcome the limitations on our powers of expression by deferring to beings who are properly able to grasp these criteria. For example, in disputes about moral status we should recognize the assessments of beings who lack our cognitive and imaginative limits as, in principle, superior to our own. The question of whether we should recognize the existence of higher moral statuses becomes the question of whether beings who are sufficiently cognitively superior to us and lack our imaginative limits would recognize the existence of such statuses.

Our deference to beings who lack our imaginative and intellectual limits resembles that which moderately talented students of mathematics grant to those whose mathematical skills are manifestly superior to their own. If you believe in the sincerity and superior mathematical skills of an interlocutor, you should sometimes believe her even when she presents conclusions about mathematics that appear unlikely. For example, it continues to seem absurd to me that 0.99 recurring could be identical to 1. But my deference to the superior mathematical judgment of others leads me to believe it.

This approach makes sense according to the prominent view of moral truth defended by Michael Smith.[30] According to Smith, true moral claims are those that would be assented to under conditions of ideal rationality and ideal information. We will never achieve these conditions. Nor will cognitively enhanced beings. Smart though they may be, they cannot know every morally relevant fact and be logically incapable of error. But we should acknowledge their perspective as superior to our own. Cognitively enhanced beings should know all the morally relevant facts known by persons. In addition, they should know facts about moral status that unenhanced humans do not. There's reason to think that their conclusions are more likely to be vindicated, at the limit of rational inquiry, than are our own.

Why Sufficiently Cognitively Enhanced Beings Will (Probably) Find That Cognitive Differences between Them and Us Mark a Difference in Moral Status

So far I've argued that we should defer to the assessments of moral status offered by sincere, cognitively superior beings. We should acknowledge them as better informed and better able to reason about relevant moral facts than we are. They lack some of our cognitive and imaginative limits and so are better able to assess claims to a higher moral status than personhood. So, how likely is it that cognitively enhanced beings will recognize the existence of higher moral statuses?

In the remainder of this section, I advance two claims. First, there is no upper limit to the capacities that determine moral status. Second, it is likely that cognitively enhanced beings would recognize a higher moral status somewhere in the spectrum of capacities higher than human persons.

Consider the following examples of bounded and unbounded capacities. There's an upper limit on the capacity to speak the English language. It's possible to specify perfect knowledge of English. This might involve knowing all of the vocabulary items and rules of grammar that are properly part of the language. It's unlikely that any current or past speaker of English has or had perfect knowledge of his or her language. But it is possible, at least in principle, to have knowledge of the English language that could not be bettered. There is a finitely large community of speakers of the language, and one could not do better than knowing everything about the language that these speakers collectively know.

There is, in contrast, no limit on possible improvements of the capacity for mathematics. It's likely that there is a very large but finite collection of mathematical truths knowable by humans. But there are almost certainly truths beyond these. There's no reason to believe that the limits of mathematics must be tied to the limits of human understanding.

Knowledge of practical reasoning seems more like knowledge about mathematics than like knowledge of the English language. It differs from knowledge of English in making no indispensable reference to principles or ideas known by a community. It's something that can always be improved. There is no reason to believe in preset limits on logical reasoning, the power of abstraction, memory, or any other of the abilities that jointly constitute practical reasoning.

Two Hypotheses about Higher Moral Statuses

Here are two hypotheses about moral enhancement that are logically compatible with the fact that powers constitutive of practical reasoning can always be improved.

Hypothesis 1   There is *some degree* of improvement of capacities constitutive of status that cognitively superior beings would recognize as creating a moral status higher than personhood.

Hypothesis 2   There is *no degree* of improvement of these capacities that cognitively superior beings would recognize as creating a moral status higher than personhood.

The available evidence should lead us to prefer hypothesis 1 to hypothesis 2. There is a limitless space of possible improvements of practical reason. There's inductive support for the notion that some degree of improvement of traits relevant to status produces a moral status superior to personhood.

Many philosophers acknowledge at least three different moral statuses up to and including persons. Inanimate objects such as rocks possess zero moral status. They are properly counted as possessing a moral status rather than lacking one. An example of something that fails to have a moral status is the concept of roundness. If presented with a dilemma in which you were required to sacrifice either a person or the concept of roundness you would have difficulty in establishing what was being requested of you. Those who seek moral trade-offs between objects that fall into the category of objects with zero status and objects that belong to higher categories receive clear advice on which objects should be favored. Rocks should be sacrificed ahead of persons. Then there are sentient nonpersons. These include sheep, cats, and dogs. They count morally, in their own right, but to a lesser extent than persons, members of the third moral status. You would be making a moral mistake if, all else equal, you rescued a rock from a burning building, deliberately leaving a cat to be consumed by the flames. You would also be mistaken if you chose to rescue the cat rather than a person.

Consider now the vast expanse of possible improvements to the human capacity for practical reason. Given the existence of three distinct moral statuses in the range of mental powers of which we currently have direct experience, it seems unlikely that no moral statuses higher than personhood could occupy this expanse. Our modest cognitive powers mean that we won't understand exactly how enhanced cognitive powers would grant a higher moral status. Remember, however, that in these matters, we're deferring to beings with understanding about morality superior to ours.

This is an inductive argument. It has some of the limitations common to inductive arguments with small evidential bases. Compare it with another inductive reasoning that seeks to reach beyond the limits of human experience. Scientists engaged in SETI (the Search for Extraterrestrial Intelligence) are searching for signs of intelligence originating from distant stars.[31] They're also interested in determining how likely it is, given what we've observed so far, that this search will be successful. We have direct evidence for intelligent life only on one planet—our own. On the other hand, the universe is vast, containing many billions of unobserved planets. Intelligent life evolved on Earth, so it's unlikely that there could be a law of nature preventing its evolution elsewhere.

The inductive argument for intelligent life beyond Earth is comparatively weak because it moves from a single observation. There's only one planet in the universe that we know to have intelligent life. The inductive argument for higher moral statuses resembles it in moving beyond evidence directly available to humans. It's stronger than SETI's inductive argument by virtue of the fact that it extrapolates from three observed moral statuses.

Moral status category 1   The zero moral status possessed by rocks.

Moral status category 2   The moral status possessed by sentient nonpersons such as sheep and toads.

Moral status category 3   The moral status possessed by persons.

These observations of moral status make it reasonable to believe in the existence of moral status category 4 that includes post-persons.

I propose that our observations of moral statuses make it likely that beings lacking our cognitive limits will recognize moral statuses superior to personhood. In the next chapter, I describe some dangerous implications of degrees of enhancement that risk producing post-persons.

9   Why Moral Status Enhancement Is a Morally Bad Thing

In what follows, I present a moral argument for avoiding the creation of post-persons. Degrees of cognitive enhancement that risk moral status enhancement should, by implication, also be avoided. This argument points to bad consequences of moral status enhancement. These consequences are not certain. They are, however, sufficiently probable and bad to justify limiting cognitive enhancement. In this chapter, I style post-persons as especially morally needy beings. The enhancement of their moral status means that their many needs should take precedence over our own. A predictable consequence is that the needs of mere persons will go unmet. We are subject to no obligation to create post-persons in the first place. We can and should avoid creating their morally weighty needs by avoiding creating them.

How should we think about the bad, but not inevitable, consequences of moral enhancement? The argument I present resembles the widely discussed consequentialist argument for reducing our emissions of greenhouse gases. This argument draws support from models of the climate that attribute some significant part of global warming to human causes. This model predicts that continuing production of greenhouse gases at current levels will have disastrous consequences for the planet's human and nonhuman inhabitants. Although not certain, these bad consequences are both sufficiently bad and sufficiently probable to justify reducing greenhouse gas production. This chapter's argument against moral status enhancement pretends certainty neither in respect of the possibility of moral status enhancement nor in respect of its bad consequences. I argue that the bad consequences are, in moral terms, so bad that a moderate probability of their occurrence makes it immoral and imprudent to not seek to prevent them.

The key objection against further anthropogenic climate change is not that no one will benefit. Technologies that produce greenhouse gases benefit many—they provide employment and returns on investments. Rather, it's that there are possible or probable consequences of climate change so bad that these jobs, profits, and other good effects do not compensate for them. It's possible that these benefits will keep their recipients safe in a warmer world. The vast majority of people will not be so fortunate, however. I argue that the consequences of moral status enhancement conform to this pattern. Any benefits received by recipients of moral enhancement do not, in moral terms, make up for the costs imposed on others.

Some Assumptions

The consequentialist argument against moral status enhancement makes certain assumptions. A first assumption is that the availability of technologies that enhance moral status will lead to mixed societies.

A *mixed society* contains both mere persons and post-persons.

Some persons will take advantage of status-enhancing technologies. Others will not, either because they reject the degree of enhancement that would enhance status or because they do not have access to status-enhancing technologies. Perhaps the technologies are very expensive. I assume an orderly transition to mixed societies.

An *orderly transition* to a new social arrangement, such as a mixed society, occurs without high levels of directly treating citizens in ways incompatible with their moral status. Supreme emergencies will be rare—no more common than they are today.

Examples of treating a person in ways that are incompatible with their moral status include murdering, enslaving, or torturing them. In an orderly transition there may be some supreme emergencies, natural or human-made disasters that justify murdering, enslaving, or torturing persons. But they will be rare. Supreme emergencies will not be much more common than they are today. It's difficult to say with any precision what might count as a high level of treating citizens in ways that are incompatible with their moral status. The level will be inferior to those resulting from the making of slave-holding societies in Europe and the Americas. It will be far inferior to that suffered by mere persons in the *Terminator* movies in which killer robots detonate nuclear weapons and set about ruthlessly eliminating the human survivors.

Orderly transitions are absent from many visions of the future that genetic and cybernetic enhancement will bring. Our cinemas present many dystopian visions of the future in which the moral status of human persons is even more egregiously offended against than in the transatlantic slave trade. Injustices that result from the separation of society into genetically enhanced haves and unenhanced have-nots are a concern of Francis Fukuyama, who worries about a future in which enhanced humans enslave humans.[1] The assumption of orderly transition supposes that degrees of moral status are recognized and respected. This chapter makes a case against degrees of enhancement that alter moral status, so it is not an inappropriate assumption for me. Obviously, if there is a disorderly transition to a mixed society then there will be many instances of persons being treated in ways incompatible with their moral status. I argue that even an orderly transition to a mixed society will cause many abuses of mere persons.

I make my assumption of an orderly transition for the sake of argument, but there may be good reason to think that it may occur. Ingmar Persson and Julian Savulescu propose that we undertake systematic moral disposition enhancement.[2] We would enhance our powers of empathy and cooperation. Had such enhancements been available and widely used in the European age of exploration, it's possible that the world would have been spared the moral calamity of the transatlantic slave trade. Enhancers of moral dispositions are too late to prevent the slave trade, but they may be able to ensure an orderly transition to mixed societies.

In what follows, I make two claims. If others become post-persons while you remain a mere person, then your moral status does not change. You do nevertheless experience a loss of relative moral status. Mere persons once possessed the highest moral status of all known beings. There are now post-persons to place ahead of you. I contend that the loss in relative moral status experienced by mere persons is likely to expose you and other mere persons to significant harm. Even on the assumption of an orderly transition, many of these harms are not justified. This may seem to contradict the assumption of an orderly transition, which supposes that people tend not to be directly treated in ways incompatible with their moral status. The term "directly" is important here. The decisions that directly cause harm may be justified, whereas decisions that are more temporally distant causes are unjustified. We should distinguish proximal from distal causes of harm. If there is an orderly transition to a mixed society, then most of the decisions that proximally cause harm are justified. This leaves open the possibility that decisions earlier in the causal sequence—more distal causes of harm—are mostly immoral. These immoral decisions anticipate later morally justified decisions to inflict harm on unenhanced humans. I will argue that decisions to create or risk creating beings with enhanced moral status are unjustified.

Why a Change in Relative Moral Status Is Likely to Lead to Significant Harms for Human Mere Persons

For some idea of the circumstances under which mere persons might suffer harms, we should begin with McMahan's discussion of supreme emergencies.[3] These rare circumstances permit the sacrifice of individuals with the highest moral status to save a (much) greater number of individuals with that same status. McMahan surmises that mere persons are likely to be better candidates for sacrifice in supreme emergencies than are beings with a moral status superior to persons. If supreme emergencies remain rare, then this may not be such a big loss. Perhaps mere persons will be adequately compensated by increases in productivity and the expansion of scientific knowledge brought about by the cognitive enhancements that enhance the moral status of other citizens.

The chief danger for mere persons lies in a foreseeable expansion of the range of cases for which they may be justifiably sacrificed. Consider another category of circumstances in which mere persons might be sacrificed. In supreme emergencies, morally significant beings are sacrificed to prevent significant harms. *Supreme opportunities* arise in respect of potential significant benefits best secured by sacrificing morally considerable beings.

Situations in which we might sacrifice morally considerable beings to secure benefits arise more frequently than situations in which we might sacrifice morally considerable beings to prevent emergencies. Suppose that emergencies are rare. Humans routinely seek benefits by sacrificing morally valuable individuals. The sacrifices of less morally considerable individuals are justified by reference to benefits produced for more morally valuable individuals. We conduct painful and lethal medical experiments on sentient nonpersons to procure important information about human diseases. We insist that suffering be minimized and not be inflicted for frivolous reasons. But we nevertheless permit it, pointing to the lower status of the beings to be sacrificed and the significant benefits conferred on beings with higher status.

Could the promise of benefits for post-persons justify the sacrifice of mere persons? One difference between supreme emergencies and supreme opportunities suggests that the answer to this question may be no.

Suppose that we limit our attention to supreme emergencies and opportunities that involve only mere persons. Some supreme emergencies permit the sacrifice of persons: it would be morally acceptable to sacrifice a small number of nuclear power plant workers to prevent its reactor from going critical and destroying a local town. But supreme opportunities involving only persons seem not to permit the sacrifice of persons. Consider the example of possible supreme benefits arising from medical research. There are strict rules restricting what can be done to human subjects in medical experiments. Such restrictions seem appropriate even when we believe that great benefits could result from a more permissive attitude. Imagine what might be learned about treating cancer in humans if we had a captive population of humans on whom to swiftly test any potentially promising therapy without the complication of consent forms or concerns about the consequences of killing a few experimental subjects. It seems morally repugnant to act on this opportunity.

What should we take from the asymmetry between supreme emergencies and supreme opportunities? Supreme emergencies seem to permit the sacrifice of persons to prevent large numbers of persons from suffering significant harms, whereas supreme opportunities seem not to permit persons to be sacrificed to bring great benefits.

There are two ways in which we can explain why mere persons cannot be sacrificed in pursuit of benefits for other mere persons. The first explanation points to a nonrelational property of persons. If we assume Buchanan's approach to persons, we would say that persons cannot be sacrificed to produce great benefits because they are capable of practical reason. This interpretation should transfer to situations involving post-persons. The act of creating post-persons would not, in itself, render mere persons incapable of practical reason. It makes no changes to the properties that prevent their sacrifice when presented with supreme opportunities.

According to a second interpretation of the prohibition, a relational property of persons prevents their sacrifice. They have the highest moral status of all involved beings. On this second interpretation, the act of creating post-persons would, in itself, dislodge mere persons from the role of having the highest moral status. Their moral demotion would render them eligible for sacrifice to provide significant benefits for post-persons.

The second interpretation has inductive support. Other gaps in moral status of which we are aware seem to permit the sacrifice of lower-status beings to benefit their moral betters. It is morally permissible to sacrifice objects with zero moral status to produce benefits for sentient nonpersons—we find it acceptable to feed carrots to rabbits, for example. It is morally permitted to sacrifice sentient nonpersons to produce benefits for persons—we find it acceptable to conduct painful and lethal experiments on rhesus monkeys to find better treatments for serious diseases suffered by human persons. These permissions provide inductive support for a permission to sacrifice mere persons to benefit post-persons.

The previous paragraph should not be interpreted as offering or assuming a blanket justification for harms inflicted by humans on sentient nonpersons. The vast majority of harms currently inflicted on sentient nonpersons may be without justification. I assume only that some harms inflicted on nonpersons are justified. I will have more to say on this below.

The first interpretation finds some support in Kantian talk of the incomparable moral value of persons. If post-persons are possible, then there could be beings whose value is not only comparable to that of mere persons, but superior to it. We could, however, grant all of Kant's normative conclusions about the situations that most interested him. In situations involving only mere persons, one would be making a serious mistake if one proposed that benefits to mere persons could justify the deliberate sacrifice of another mere person. We are required to emend Kantian views about the incomparable value of (mere) persons only in circumstances involving post-persons. Such scenarios were irrelevant to Kant's purpose of an account of morality fit for the world as it presented to him.

McMahan illustrates the difference in moral status between post-persons and mere persons by means of an example involving a supreme emergency. He proposes that mere persons might be more eligible for sacrifice in such cases than post-persons. In limiting his discussion to supreme emergencies, McMahan does not do more than is required to establish the possible existence of beings with a status higher than mere persons. His argument need not be read as implying that supreme emergencies are the *only* circumstances in which differences between mere persons and post-persons become apparent. It is perfectly compatible with McMahan's view that the difference in status licenses other forms of discrimination.

Thomas Douglas has suggested to me (in personal communication) that if the differences in status are real but small then morality might require that mere persons be distinguished from post-persons *only* in supreme emergencies. There are good inductive grounds for thinking that the differences in moral status between mere persons and post-persons will be quite significant. The gaps between the status of nonsentient things and sentient nonpersons and between sentient nonpersons and persons make big differences to permissible or required treatment. The differences are certainly not limited to supreme emergencies.

Remember that we are treating mere personhood and post-personhood as different weak thresholds and not just different points on a graph registering continuous improvements of a morally relevant scalar property. If different points on a line registering continuous moral status improvement are close to one another, it is reasonable to think that moral requirements and permissions corresponding to them may be similar. Weak thresholds combine a comparatively wide range of different degrees of a morally relevant attribute. As a consequence, there's likely to be a big difference between requirements and permissions appropriate for each weak moral threshold.

Consider the example I used to introduce the idea of weak thresholds: weak thresholds separate students enrolled in introductory, intermediate, and advanced language classes. There would be a small difference between these classes if only a small number of vocabulary items or grammatical principles taught in the higher class were not taught in the lower class. But this is unlikely to be the case. The instruction properly offered in introductory classes differs quite markedly from the instruction offered in intermediate language classes. This is so even if there is only a small difference in the language skills of the most knowledgeable student in the introductory class and the least knowledgeable student in the intermediate class.

Peter Singer has commented extensively on our inconsistent attitude toward those who do not satisfy the criteria for personhood.[4] We permit nonhuman animals to be farmed and experimented on. We would never dream of allowing humans with similar cognitive abilities—for example, infants, the severely cognitive disabled—to be treated in this fashion. Morally unjustified speciesist preferences may tend to work against mere persons in the kinds of scenarios we are imagining. The enhancements that turn persons into post-persons may weaken these bonds of empathy. Suppose that their inferior status justifies the factory farming of and medical experimentation on nonhuman animals. It would also justify this treatment of human nonpersons. We now choose to make exceptions for nonpersons who are members of our biological species. If post-persons feel no kinship with mere persons they will probably cease to recognize us as worthy of any speciesist exemptions.

Douglas is wary of inferring too much from benefits produced for humans that result from harms inflicted on lower-status beings.[5] He makes the point that much harm we inflict on animals is in fact morally unjustified. The factory farming of animals is arguably immoral. Much experimentation on animals inflicts significant harms on them to produce inconsequential benefits for humans. Perhaps it is reasonable to suppose that post-persons will take advantage of their great power to inflict unjustified harms on human persons. Such harms are a standard feature of science fiction depictions of relations between the enhanced and unenhanced. But these scenarios are not compatible with the assumption of orderly transition that I presented at the beginning of this chapter. An orderly transition to a mixed society occurs without high levels of directly treating citizens in ways incompatible with their moral status. I offer this as a concession to the advocates of radical enhancement. I claim that mere persons will suffer significant harms even if there is an entirely orderly transition to a mixed society. Some medical experimentation inflicted on nonpersons is not justified, but some surely is. As a person with insulin-dependent diabetes I feel grateful for the dogs who died in the experiments of the Canadian researchers Frederick Banting and Charles Best that led to the discovery of insulin. Opponents of animal testing sometimes like to say that we have ways of conducting medical research that do not involve animal testing. It's true that had Banting and Best not experimented on dogs, we would have discovered insulin by other means. But such arguments ignore the human death toll that would have resulted from forgoing the most direct method, which involves animal experimentation. In short, the existence of human persons leads sentient nonpersons to suffer much justified harm. It's reasonable to suppose that the existence of post-persons will lead mere persons to suffer many harms that post-persons will view as morally justified.

Why Post-Persons Will Probably Identify Many Supreme Opportunities Requiring the Sacrifice of Mere Persons

How does the addition of supreme opportunities add to the likely burdens of mere persons in a society some of whose members are post-persons? Suppose that the transition to a mixed society is orderly. Supreme opportunities will permit mere persons to be sacrificed to provide significant benefits for post-persons. Just as human persons are morally entitled to sacrifice sentient nonpersons in pursuit of better treatments for serious human diseases, so too post-persons may be entitled to sacrifice mere persons to gain a better understanding of ailments afflicting them. Human medical researchers use monkeys because their relatedness to us makes them a useful model for human disease. Suppose that diseases afflict post-persons. The emergence of post-persons from human mere persons may make mere persons ideal subjects for medical experiments.

This is but one example of a possible use that post-persons may make of mere persons. The cognitive enhancement that may turn mere persons into post-persons is likely to generate uses for their human person ancestors that we cannot identify. The enhancement of cognitive powers that occurred with the evolution of humans from apelike ancestors has created beneficial uses for many parts of the environment for which apes have no use. Super-intelligent post-persons are likely to find beneficial uses of parts of their environment that we cannot think of. Some of those parts of their environment could include human brains and bodies.

We shouldn't presume too much insight into the designs of beings with radically enhanced intellects. The futurist Ray Kurzweil has a suggestion that should scare mere persons.[6] Kurzweil predicts that advances in information technologies will soon set off a progression of increasingly powerful cognitive enhancements. One way to enhance the processing power of our minds is to physically expand them. Cognitively enhanced beings would do the same thing to their minds that computer engineers do when they add more transistors to a computer. Kurzweil predicts a future in which the minds of enhanced humans colonize the universe. Every bit of matter and energy will become the substrate of, and fuel for, thought. This could include the matter and energy that constitute the brains and bodies of human mere persons.

I conclude that it's reasonable to think that the creation of post-persons will leave the mere persons more likely to suffer significant harms.

What Complaint Can Mere Persons Make about the Harms They Suffer in Mixed Societies?

Up until this point I have argued that post-persons are likely to inflict many harms on mere persons. What I have yet to establish is that these harms would be morally bad. Indeed, the assumption of an orderly transition seems to have granted that harms inflicted on mere persons by post-persons will tend to be morally justified. We may not enjoy experiencing these harms, but post-persons will have a ready response to any complaints that we might make. In his defense of moral status enhancement, Douglas explores some of the defenses that post-persons might offer.

One complaint about moral status enhancement is that it has generated inequality. The transition to a mixed society takes a group of citizens with equal status and boosts the status of some while leaving the status of others unchanged. We arrive at a less equal distribution of moral immunity to harm and moral eligibility for benefit. Douglas doubts that this manner of equality really matters. According to him, it's not something upon which we place value. This is because moral immunity is a normative notion. Douglas says that "one cannot maintain that inequality of immunity exists without that inequality being, in a sense, appropriate or justified. ... One being has greater immunity to permissible harm than another just in case it enjoys stronger moral claims against certain kinds of harm, and that cannot be so if the difference in the strength of those claims is unjustified."[7] One cannot have a reduced *moral* immunity unless one morally deserves it. Douglas offers as an example the property of culpability. Differences in culpability give rise to differences in liability to punishment. But it is entirely appropriate that they do so. Differences in culpability not only give rise to differences in liability to punishment, they *justify* them.

The normativity of moral status permits an analogous answer. According to Douglas:

If status enhancements give rise to further inequalities in immunity, those new inequalities must also be, in a sense, justified. They are justified by the very differences in moral status which produce them. Since immunity to permissible harm is a normative notion, one being cannot have greater immunity than another unless the difference is justified. So a new difference in moral status could not give rise to a new inequality in immunity unless it also justified it.[8]

Douglas offers for comparison the harms that the lower status of chickens exposes them to. Humans are permitted to prevent harm to themselves by redirecting that harm to chickens. It's precisely this manner of difference that permits the post-person to inflict harm on the mere person.

I think that Douglas is right to exempt from moral condemnation the particular decisions of post-persons to inflict harm. But this does not mean that no moral criticism can be offered. Consider the moral condemnation we might offer of a terrorist who plants and detonates a bomb in a crowded shopping mall. The detonating of the bomb is preceded by many decisions without which the atrocity could not have happened. Let us suppose that among these decisions is the decision of a clerk to sell fertilizer to the bomber. We can presume that the clerk would defend her decision—she had no grounds for suspecting the bomber who did a good job of passing himself off as a gardener in need of agricultural necessities. But the fact that the clerk is innocent does not prevent there being other decisions—namely, the decisions to manufacture, plant, and detonate the bomb—that warrant condemnation. Moral criticism of an act is appropriate when there is *one such decision* in the causal chain leading up to it. The task for those who offer criticism is to locate this morally weak link. In the case of the moral status enhancement we find the morally weak link well before the decisions of post-persons to use the difference in status between them and mere persons to justify inflicting suffering.

Consider an analogous issue concerning global climate change. Some commentators think that a loss of habitable and cultivable land will lead to widespread unrest as groups fight for control over diminishing supplies of life-sustaining resources. But this is not a necessary consequence. It's possible that there will be an entirely orderly transition from circumstances of relative plenty to circumstances of relative scarcity; decisions about how to distribute the diminished supplies of vital resources will be made in accordance with all relevant moral criteria. The climate crisis is one of the prompts for Savulescu and Persson's case for moral disposition enhancement.[9] If there is something like an orderly transition to circumstances with reduced life-sustaining resources, then complaints from those deprived of food and shelter should not be directed at the decisions that were the proximal causes of their deaths. According to the assumption of orderly transition, these will tend to be justified—the decisions to deny them food and shelter will be informed by morally relevant criteria. Moral criticisms should instead be directed at distal causes—the earlier choices that permitted circumstances such as these to arise. They will criticize our decision to continue emitting high levels of greenhouse gases when the detrimental effects of doing so were apparent.

If there is an orderly transition to mixed societies, then post-persons will only rarely abuse their power over mere persons. They will, in general, treat mere persons exactly as they deserve. I argue that this will entail significant sacrifices by mere persons. Mere persons may not complain about the decisions by post-persons to impose those sacrifices. The difference in moral status between post-persons and mere persons makes those sacrifices legitimate. But mere persons can complain about choices that anticipate the morally correct choices by post-persons—namely, the choice to create beings with enhanced moral status.

Douglas is aware that injustice could attach at the point of the creation of differentials in moral status rather than at the point at which these differentials are used to justify inflicting harms. He describes one unjust way of distributing status enhancements:

Suppose that the technologies enabling status enhancement are monopolized by a group of scientists who ensure that all members of a certain racial or socio-economic group are prevented from accessing these technologies. As a result, though many others would like to, only members of the privileged group(s) actually undergo status enhancements. In this case, it seems plausible that we wind up with an unjust distribution of mental capacity across different beings.[10]

Douglas contends that we can defend moral status enhancement itself by identifying ways in which differentials are created that do not stem from these violations of distributive justice:

Suppose, for example, that access to status enhancements is decided via a lottery procedure to which all agree in advance. Now, it seems doubtful whether the resulting distribution of mental capacity would be unjust or otherwise disvaluable. Or suppose status enhancements were made readily available to all, but that only some chose to use them, with others preferring to remain mere persons.[11]

Douglas's suggestion that we might correct injustice by modifying distributive schemes seems to suggest that any moral complaints are not properly directed at the status enhancements themselves. We can eliminate the injustice without making any change to the status enhancement. The complaint would be like that of someone who challenges the distributive arrangements of a nation's health care system. This complaint allows that health care is, itself, a good thing.

Douglas's discussion of distributive justice does not adequately respond to concerns about moral status enhancement. For example, one may decide that Russian roulette is a method of deciding who among a group of six individuals should be shot that is morally preferable to some alternatives. For example, it eliminates a distributive injustice attaching to an arrangement in which the individual who happens to own the gun decides who should be shot. But correction of this variety of injustice does not demonstrate the moral acceptability of an arrangement according to which one of the six must be shot. If there were a way to avoid this scenario, then it is morally required to do so. I suspect that similar points apply to moral status enhancement. Its consequences are so significant that it should be avoided. We do not adequately respond to them by identifying a way of distributing resulting suffering that is free of a particular distributive injustice.

Why a Loss of Relative Status Is Unlikely to Be Adequately Compensated

Perhaps the burden of a decline in relative moral status can be rectified by some act of compensation. Suppose that the creation of post-persons increases the frequency and severity of harms suffered by mere persons. These forecast harms might be justified if we believed that sufficient benefits accrued to those who will suffer them. This manner of justification is offered for the upheavals brought by the introduction of new technologies into an economy. Textile workers lost their livelihoods as a result of the industrial revolution but ended up better off as a result of the creation of new more remunerative jobs. If they didn't, then their children did. Or so the story goes.

I suspect that mere persons are likely to be less eligible for benefits than were textile workers whose jobs were destroyed by the industrial revolution. Mere persons are unlikely to be good candidates for benefits resulting from cognitive enhancement in circumstances that include post-persons. The enhanced moral status of post-persons makes them systematically morally preferable candidates for benefit. Consider the medical benefits currently derived from inflicting harm on sentient nonpersons. It would be viewed as a serious moral mistake to place a sentient nonperson sufferer ahead of a human person sufferer on a waiting list. This is because the higher status of the person makes addressing her needs more important than addressing the needs of a nonperson. The difference in moral status between post-persons and mere persons is likely to result in a similar prioritization. The significant needs of post-persons will take priority over the significant needs of mere persons.

Consider an example of how differentials in moral status trump concerns about which beings might have suffered harms in the generation of a benefit. Animals are routinely experimented on in medical research. This research has the goal of finding better therapies for human diseases. Some of the diseases addressed by medical researchers are suffered by species that provide the experimental subjects. It seems that the fact that the members of a species are candidates for medical experiments does little to make up for the reduced entitlement brought by their moral status. Rhesus monkeys may perform valuable services in research on Parkinson's disease. But we would consider using the expensive therapies that such research produces on rhesus monkey sufferers of Parkinson's only after all human patients have been adequately treated. We would appeal to facts about relative moral status to vigorously challenge a doctor who sought to place a rhesus monkey patient on a waiting list for an expensive treatment ahead of a human patient.

In this chapter and the one that preceded it, I've argued for two claims. First, it is likely that some degree of moral enhancement will enhance moral status. There are good inductive grounds for thinking that it will bring into existence post-persons—beings with a status superior to mere persons. Second, the creation of post-persons would be a morally bad thing for mere persons. It is likely to impose significant penalties on them. The fact that there is no moral obligation to create post-persons means that we should not. The enhancement of moral status is not something that can be aimed at in the way that we might aim at the enhancement of mathematical skills or of our ability to learn foreign languages. In the latter case we know what would count as success. The insidious thing about moral status enhancement is that it is likely to occur as an unintended by-product of enhancement directed at human cognitive abilities. It imposes a significant undeserved penalty on those who have not received the enhancements that bring it about.

10 A Technological Yet Truly Human Future—as Depicted in *Star Trek*

This book has presented an ideal of truly human enhancement that combines an endorsement of moderate enhancement with a rejection of radical enhancement.

Chapter 7 contained an argument in favor of moderate human enhancement—the improvement of significant attributes and abilities to levels *within or close to* what is currently possible for human beings. This degree of enhancement tends not to involve the imaginative obstacles erected by radical enhancement. To reject moderate human enhancement is to reject much of what we rightly consider good child-rearing. Parents of gifted children do not deliberately terminate their educations once they have achieved levels of performance deemed normal for human beings. It is outrageous to suppose that they should. Most opponents of moderate enhancement focus on the means by which these achievements are procured. They contend that there's something wrong with enhancement targeted at humans' hereditary material. I argued that the distinction between genetic and environmental enhancement lacks moral significance. Valid concerns about genetic enhancement are more properly directed at genetic enhancements of too great a degree. These concerns apply with equal force to environmental enhancements of too great a degree. We should accept some moderate human enhancements.

My criticisms of radical enhancement fell into two categories. I argued that some radical enhancements are prudentially irrational. They seem, at the outset, to offer a great deal. But they give us experiences and achievements that we properly view as less valuable than those that fill unenhanced human lives. I also argued that radical enhancement is immoral. In chapter 6, I addressed the actions that are likely to be required to make radical life extension a reality. These will probably involve subjecting healthy poor people to dangerous medical experiments. In chapters 8 and 9, I presented a rather different, more abstract moral criticism. Radical cognitive enhancement is likely to create beings with a moral status superior to persons. The existence of these morally needy post-persons will lead significant, uncompensated harms to be inflicted on human mere persons.

I want now to shift focus from details of argument to the ideal of truly human enhancement as a positive vision of the future of our species. I propose that it embodies an especially attractive conception of where we might collectively be headed and our future relationship with technology.

The opening chapter of this book explored some scary possible human futures taken from science fiction. We encountered scenarios in which aliens subject human beings to the transformative changes of body-snatching, cyberconversion, and Borg assimilation. These changes were unpleasant in a quite specific and insidious way. Not only did they alter us in ways that we viewed as significant and distressing, but they altered the way we evaluated the changes. We became beings who valued changes that we formerly rightly rejected. I contended that radical enhancement—the improvement of significant attributes and abilities to levels that *greatly exceed* what is currently possible for human beings—is a form of transformative change. It shares some of the distressing features of body-snatching, cyberconversion, and Borg assimilation. It is rationally irreversible. It leads us to endorse changes that we were right to reject. I conclude this book with a discussion of a science fiction portrayal of a more optimistic vision of how humans might relate to and be affected by technological progress.

Consider an iconic depiction of humans interacting with powerful future technologies—the original series of *Star Trek*. The series showcases a 1960s vision of where three centuries of technological progress will take the human species. The predominantly human crew of the starship *Enterprise* is ferried from star system to star system at many times the speed of light. Upon arrival they beam down to planet surfaces by means of a technology that disintegrates their bodies, producing exact replicas at their chosen destinations. They use torpedoes charged with antimatter to pound into space dust those who wrongfully obstruct their passage. In spite of all of this, they themselves, with the half-alien exception of the Science Officer Mr. Spock, present as recognizably, and indeed unmistakably, human. It's really not too hard to imagine Captain James T. Kirk of the starship *Enterprise* maturing into the Bill Shatner of early twenty-first-century sitcoms and talk shows. He seems no more adept at responding to mental and physical challenges than would be many of the human beings who make up his television audience. We recognize many of his foibles and vulnerabilities as our own. We can almost make sense of his predilection for green-skinned alien women.

This, on the face of it, is extremely odd. For the *Star Trek* vision of our future to be true, something must have prevented the forces that turned Boeing 747s into starships, stethoscopes into medical tricorders, and pistols into phasers from turning the human crew members of the starship *Enterprise* into beings whose genetic and cybernetic enhancements make their capacities, experiences, and aspirations very different from our own. Did the screenplay writers of *Star Trek* simply omit to consider human enhancement technologies?

There are some uninteresting reasons for the apparent anachronism at the heart of *Star Trek: The Original Series*. It was made to appeal to the tastes of audiences composed of late-twentieth-century human beings. They are likely to have found earnest attempts to depict a technologically transformed twenty-third-century humanity as emotionally uninvolving as would have readers of Richard Adams's novel *Watership Down*, had the author made the rabbits who were his central characters psychologically realistic members of the genus *Oryctolagus*.

This is not how I choose to view *Star Trek*. Rather than seeing it as a mistake or something imposed on the series producers by the tastes of their audience, we should treat the familiarly human crew of the starship *Enterprise* as an optimistic statement of a possible future for our species. In this future, we achieve all manner of technological advances. We travel to the stars. We find cures for cancer. We build powerful machines that help us to deal with the computational complexities of some presentations of the "Ultimate Question of Life, the Universe and Everything." Because we give these machine assistants well-defined assignments, we aren't completely dumbfounded by their answers, even when we know that our unaided brains could never have found them. They won't spring "42" on us. We can have all of this without consenting to become whatever technology can turn us into. We remain a technological species in the sense that we *use* technology, rather than in the sense of being a species destined to *become* technology. *Star Trek* and futuristic visions of its type serve as optimistic expressions of the idea that human beings are good enough to travel to the stars. We aren't required to transform ourselves into different kinds of beings—robots or posthumans—to be worthy of that honor. Knowingly or not, viewers and readers indulge a hope that the wonderful events of the twenty-third century be proper parts of the human story rather than mere addenda to it.

Is this too stark a depiction of the effects of enhancement technologies on our characters and values? We are unlikely to choose to abruptly turn ourselves into superhumans or cyborgs. If they do occur, such changes will occur gradually. Over several years we will apply a sequence of increasingly powerful enhancement technologies to our psyches and physiques. There will be no jarringly abrupt transformation. If the arguments of this book are correct, we are entitled to view the gradual transition as gradually eroding features of ourselves that we rightly value. Finding a significant difference between the abrupt and gradual loss of humanity is a bit like finding a big difference between the thief who would take all of your money in one dramatic heist and the thief who methodically drains your accounts over the course of a few weeks. If our distinctive human values truly matter to us, then their loss is bad even if it happens gradually and without our noticing.

We need not suppose that we will survive as we are indefinitely for our distinctive human values to be worth preserving. At some point, should our lineage not go extinct, we are almost certain to be replaced by beings different from us and with values different from our own. We should approach this collective demise in the way we think about our own personal extinction. Death is inevitable. But that is no reason to hurry it along.

This emphasis on humans, human experiences, and human achievements might sound exclusionary. This is how the transhumanist writer James Hughes presents it.[1] He accuses opponents of radical enhancement of "human racism." According to him, rejecting the radical enhancement of your cognitive capacities exhibits the same bigotry that permits humans to inflict great suffering on animals. Their lack of humanity does not justify such acts. It would be equally wrong to use the possible lack of humanity of radically enhanced beings to justify causing them to suffer. We must distinguish the moral view that (mistakenly) permits us to inflict severe harm on the grounds of a lack of humanity from the prudential view about what it is good for human beings to become. One can certainly insist that we give significant weight to the suffering of chimpanzees while at the same time holding that it is (prudentially) bad for a human being to undergo a transformative change into a chimpanzee. This is so even if the human would become a very contented chimpanzee. The recognition that a radically enhanced posthuman is a bad thing for a human to become should coexist with the acknowledgment of the moral worth of any posthumans who happen to come into existence.

One way to think about our interest in a human future is to compare it with the distinctive interest we take in the human past. History is fascinating whether or not it involves human beings. But historical events involving human actors have a special significance for us. Consider the debate among anthropologists over when in our biological lineage the first human being appeared.[2] We know that human beings evolved from apelike tree-climbing species. We know that a number of changes separate these apelike creatures from ourselves. We are bipedal, have opposable thumbs, use tools, build shelters, have spoken language, and so on. Our distant ancestors lack all or many of these traits. So, who was the first ancestor to combine sufficiently many of these improvements to properly count as human, and when did this happen? Considered on its scientific merits, this question is not particularly interesting. There are likely to be different explanations for each of these traits. There's no *a priori* reason to think that they must have evolved as some specifically human evolutionary package. For example, a good scientific account of the emergence of opposable thumbs may have little to do with a good scientific account of the emergence of spoken language.

I propose that speculations about the first humans engage another, nonscientific interest in the past. The question of when the first human emerged invites us to engage in a certain imaginative identification. We can certainly seek to imagine ourselves as a triceratops that has just caught the scent of a tyrannosaur. We can seek to imagine ourselves as a glacier retreating in the face of rising global temperatures. These can be emotionally rewarding imaginative exercises. But they are fanciful. They are not veridical. They require a make-believe projection of human psychological states onto dinosaurs and glaciers. Questions about the first human invite human beings to speculate about how far they can veridically project themselves into the past. The circumstances of the first humans were very different from our own. Their experiences would differ from our own in various ways. But they would have been human experiences. Inhabitants of Papua New Guinea recognize Icelanders as having different experiences. But they correctly recognize these as *human* experiences. They are experiences that they might have had, had the circumstances of their births been different. How far is it legitimate to project ourselves into the past? If we go back 300,000 years, do we find the most temporally distant members of the human community? An answer to this question engages a distinctively human interest in the past.

We have the same interest in the future. We can veridically project ourselves into the possible future populated by Captain Kirks and Lieutenant Uhuras. We expect that the worlds of these future humans will be very different from our own. But we can look upon their experiences as ones that we might have had, had our births been postdated by a few hundred years. The transformative change of radical enhancement brings a future populated by beings with whom we can identify only nonveridically. Our attempts to imaginatively engage with them will involve the same kinds of error as our attempts to engage with triceratops and glaciers. We have an interest in the human story. We want to know when that human story began. We would like it to continue. The human story is a complex, multibranched narrative that links the universality of human experiences. We know that certain kinds of environmental catastrophe have the power to end the human story. We should also appreciate that the technologies of enhancement have that same power.

Notes

1 Radical Human Enhancement as a Transformative Change

1. For this definition, see Nicholas Agar, *Humanity's End: Why We Should Reject Radical Enhancement* (Cambridge, MA: MIT Press, 2010), ch. 1. See also Nick Bostrom's definition of a posthuman capacity in Nick Bostrom, "Why I Want to Be a Posthuman When I Grow Up," in *Medical Enhancement and Posthumanity*, ed. Bert Gordijn and Ruth Chadwick (Dordrecht: Springer, 2009), 108.

2. Thanks to Erik Parens for suggesting this name to me.

3. Ray Kurzweil, *The Singularity Is Near: When Humans Transcend Biology* (London: Penguin, 2005), 136.

4. Ibid.

5. See Rosalind Hursthouse, "Virtue Theory and Abortion," *Philosophy and Public Affairs* 20 (1991): 223–246, at 231. Hursthouse observes: "There are youthful mathematical geniuses, but rarely, if ever, youthful moral geniuses, and this tells us something significant about the sort of knowledge that moral knowledge is." I do not mean here to endorse Hursthouse's virtue ethical account, merely to echo her views about the necessary difficulty of moral evaluation.

6. Robert Nozick, *Philosophical Explanations* (Cambridge, MA: Harvard University Press, 1981), 59.

7. Two examples are the accounts of Peter Unger, *Identity, Consciousness, and Value* (New York: Oxford University Press, 1990), and Eric Olson, *The Human Animal: Personal Identity without Psychology* (Oxford: Oxford University Press, 1999).

8. George Annas, "The Man on the Moon," in *Science Fiction and Philosophy: From Time Travel to Superintelligence*, ed. Susan Schneider (Malden, MA: Wiley Blackwell, 2009), 227–240, and Francis Fukuyama, *Our Posthuman Future: Consequences of the Biotechnology Revolution* (New York: Farrar, Straus & Giroux, 2002).

9. John Stuart Mill, *The Collected Works of John Stuart Mill*, 33 vols., General Editor, John M. Robson (Toronto: University of Toronto Press, 1963–1991), vol. 10, 212.

10. For an argument that many opponents of human enhancement make irrational appeals to the status quo, see Nick Bostrom and Toby Ord, "The Reversal Test: Eliminating Status Quo Bias in Applied Ethics," *Ethics* 116 (2006): 656–679.

11. Kurzweil, *The Singularity Is Near*, 310.

2   Two Ideals of Human Enhancement

1. For defenses of the bioconservative view, see Francis Fukuyama, *Our Posthuman Future: Consequences of the Biotechnology Revolution* (New York: Farrar, Straus & Giroux, 2002); Leon Kass, "The Wisdom of Repugnance: Why We Should Ban the Cloning of Humans," *New Republic*, June 2, 1997; Leon Kass, *Life, Liberty, and the Defense of Dignity: The Challenge for Bioethics* (San Francisco: Encounter Books, 2002); Bill McKibben, *Enough: Staying Human in an Engineered Age* (New York: Times Books, 2003).

2. In his 2011 book *Beyond Humanity: The Ethics of Biomedical Enhancement* (New York: Oxford University Press, 2011, 13), Allen Buchanan characterizes his position as "anti-anti-enhancement." This nuanced take on enhancement resembles the more philosophically straightforward pro-enhancement stance in rejecting the reasoning of those who find *any* human enhancement to be morally unacceptable. But it differs from a pro-enhancement position in allowing that some human enhancements could be morally objectionable. The position to be defended in this book is anti-anti-enhancement in Buchanan's sense. I argue that some human enhancements are morally permissible. My position is also—to adapt Buchanan's language—*anti-pro-enhancement*. The book opposes degrees of enhancement endorsed by many pro-enhancers. Rather than describing a position that is logically compatible with the impermissibility of some human enhancements, I identify a feature that separates bad enhancements from good ones.

3. See, e.g., Buchanan, *Beyond Humanity*, 5; Julian Savulescu and Nick Bostrom, "Human Enhancement Ethics: The State of the Debate," in *Human Enhancement*, ed. J. Savulescu and N. Bostrom (Oxford: Oxford University Press, 2009), 3.

4. Savulescu and Bostrom, "Human Enhancement Ethics," 3.

5. See Christopher Boorse, "On the Distinction between Disease and Illness," *Philosophy and Public Affairs* 5 (1975): 49–68, and Norman Daniels, "Normal Functioning and the Treatment-Enhancement Distinction," *Cambridge Quarterly of Healthcare Ethics* 9 (2000): 309–322.

6. Hans Moravec, "When Will Computer Hardware Match the Human Brain?" *Journal of Evolution and Technology* 1 (1998), http://www.jetpress.org/volume1/moravec .htm.

7. Nick Bostrom, "Why I Want to Be a Posthuman When I Grow Up," in *Medical Enhancement and Posthumanity*, ed. Bert Gordijn and Ruth Chadwick (Dordrecht: Springer, 2009), 108–109.

8. Margaret Talbot, "Brain Gain: The Underground World of 'Neuroenhancing' Drugs," *New Yorker* 85, no. 11 (April 27, 2009): 32–43.

9. "The Transhumanist FAQ," collated by Nick Bostrom and available at http://www.transhumanism.org/resources/FAQv21.pdf.

10. Ray Kurzweil, *The Singularity Is Near: When Humans Transcend Biology* (London: Penguin, 2005), 136.

11. Ibid., 7.

12. Ibid., 57.

13. Ibid., 61.

14. Ibid., 71.

15. Ibid., 73.

16. Ibid., 80.

17. Max More, "A Letter to Mother Nature," http://www.maxmore.com/mother.htm.

18. See John O'Neill, "The Varieties of Intrinsic Value," *Monist* 75 (1992): 119–137.

19. Alasdair MacIntyre, *After Virtue: A Study in Moral Theory* (Notre Dame: University of Notre Dame Press, 1984), 188.

20. Ibid., 188–189.

3 What Interest Do We Have in Superhuman Feats?

1. Robert Freitas, Jr., "A Mechanical Artificial Red Cell: Exploratory Design in Medical Nanotechnology," http://www.foresight.org/Nanomedicine/Respirocytes.html.

2. Robert Freitas, Jr., "Robots in the Bloodstream: The Promise of Nanomedicine," http://www.kurzweilai.net/robots-in-the-bloodstream-the-promise-of-nanomedicine.

3. Bill McKibben, *Enough: Staying Human in an Engineered Age* (New York: Times Books, 2003), 6–7.

4. For this argument, see Nicholas Agar, "Sport, Simulation, and EPO," in *The Ideal of Nature*, ed. Gregory Kaebnick (Baltimore: Johns Hopkins University Press, 2011), 149–167.

5. For a useful introduction to simulation theory, see Robert Gordon, "Folk Psychology as Mental Simulation," in *The Stanford Encyclopedia of Philosophy* (fall 2009 ed.), ed. Edward N. Zalta, http://plato.stanford.edu/archives/fall2009/entries/folkpsych -simulation/.

6. See V. S. Ramachandran, *The Tell-Tale Brain: A Neuroscientist's Quest for What Makes Us Human* (New York: W. W. Norton, 2010), for an accessible recent presentation and defense of mirror neurons.

7. See, e.g., Gregory Currie and Ian Ravenscroft, *Recreative Minds: Imagination in Philosophy and Psychology* (New York: Oxford University Press, 2003), and Gregory Currie, "The Moral Psychology of Fiction," in *Art and Its Messages: Meaning, Morality, and Society*, ed. Stephen Davies (University Park, PA: University of Pennsylvania State Press, 1997).

8. Currie, "The Moral Psychology of Fiction," 56.

9. Leon Kass and Eric Cohen, "For the Love of the Game," *New Republic* (March 26, 2008): 38.

10. Ibid.

11. For an interesting discussion of Deep Blue's affectless play, see Charles Krauthammer, "Deep Blue Funk," *Time* (February 26, 1996).

12. These statistics come from the IBM website, http://www.research.ibm.com/deepblue/meet/html/d.3.2.html.

13. Rob Sparrow has discussed a further dangerous implication of internalizing enhancement. The problem of integration presents initially as a challenge to those seeking to enhance our access to certain goods by linking technologies directly to our brains and bodies. Old televisions can simply be replaced by sets that result from technological advances. Enhancements built into our brains and bodies cannot be so easily trashed. They have become part of us.

14. Kurzweil, *The Singularity Is Near*, 127.

15. Neil Levy, *Neuroethics: Challenges for the 21st Century* (Cambridge: Cambridge University Press, 2007), 29.

16. Andy Clark, *Natural-Born Cyborgs: Mind, Technologies, and the Future of Human Intelligence* (New York: Oxford University Press, 2004), 3.

4   The Threat to Human Identities from Too Much Enhancement

1. Derek Parfit, *Reasons and Persons* (Oxford: Oxford University Press, 1984).

2. Bernard Williams, "The Self and Its Future," in Bernard Williams, *Problems of the Self* (Cambridge: Cambridge University Press, 1973), 46–63.

3. Walter Glannon, "Identity, Prudential Concern, and Extended Lives," *Bioethics* 16 (2002): 266–283, at 268.

4. Ibid., 268.

5. Ibid., 279.

6. Ibid.

7. Ibid.

8. John Harris, "A Response to Walter Glannon," *Bioethics 16* (2002): 284–291, at 285.

9. For an accessible recent introduction to this view of memory see Charles Ferny-hough, *Pieces of Light: How the New Science of Memory Illuminates the Stories We Tell about our Pasts* (New York: HarperCollins, 2013).

10. Thanks to an anonymous referee for the MIT Press for this observation.

11. Nick Bostrom and Toby Ord, "The Reversal Test: Eliminating Status Quo Bias in Applied Ethics," *Ethics 116* (2006): 656–679, at 671.

12. See Ronald Dworkin, *Life's Dominion: An Argument about Abortion, Euthanasia, and Individual Freedom* (New York: Knopf, 1993), 218–229, for a persuasive argument for this conclusion.

13. Bernard Williams and J. J. C. Smart, *Utilitarianism: For and Against* (Cambridge: Cambridge University Press, 1973).

14. Kurzweil, *The Singularity Is Near*, 7.

15. Ibid., 9.

16. Ibid., 29.

5   Should We Enhance Our Cognitive Powers to Better Understand the Universe and Our Place in It?

1. Some philosophers of science find differences between scientific prediction and explanation. These are of no importance to the present discussion. I use the term "explanation" to cover both explaining and predicting.

2. Pierre-Simon Laplace, *A Philosophical Essay on Probabilities*, translated into English from the original French 6th ed. by F. W. Truscott and F. L. Emory (New York: Dover, 1951), 4. Some interpretations of quantum mechanics make this kind of knowledge about the universe an impossibility. According to Heisenberg's uncertainty principle it is impossible to know, at any given time, the exact position and momentum of certain particles. Perhaps Laplace's demon cannot know both of these facts about subatomic particles. We can suppose that if Laplace's demon is not a perfect explainer and predictor then it is as good an explainer or predictor as any being could ever be. No manner of enhancement could ever improve upon its powers of explanation or prediction. For the purposes of the discussion that follows, I ignore this complication and refer to Laplace's demon as a perfect explainer.

3. Some scholars will say that Laplace intended his demon only to be a perfect predictor and hence would not have credited it with counterfactual knowledge. He intended the demon to be a very good predictor but a very poor explainer. I make apologies to the historians but nevertheless include counterfactual knowledge in the demon's knowledge base.

4. Karl Popper, *The Logic of Scientific Discovery* (London: Routledge, 1959).

5. The label "relativist" is one that scientists and philosophers tend to resist. Among those sometimes described as scientific relativists are Paul Feyerabend and David Bloor.

6. Idealization is a central feature of many philosophical presentations of science ranging from the view of Nancy Cartwright, who combines a realistic approach to the theoretical entities of physics with an argument that the laws of physics are idealizations that deliberately misrepresent reality (see Nancy Cartwright, *How the Laws of Physics Lie* [Oxford: Oxford University Press, 1983]), to the scientific antirealism of Bas van Fraassen (see the discussion of probability assignments in section 4.4 of Bas van Fraassen, *The Scientific Image* [Oxford: Oxford University Press, 1980], section 4.4) and the perspectival realism of Ronald Giere, which posits models that approximate to reality (see Ronald Giere, *Explaining Science: A Cognitive Approach* [Chicago: University of Chicago Press, 1988]).

7. Roman Frigg and Stephan Hartmann, "Models in Science," *The Stanford Encyclopedia of Philosophy* (fall 2012 ed.), ed. Edward N. Zalta, http://plato.stanford.edu/archives/fall2012/entries/models-science/.

8. See http://education.jlab.org/qa/mathatom_04.html.

9. Thanks to an anonymous referee for convincing me to broaden my discussion of scientific idealization.

10. See Michael Strevens, *Depth: An Account of Scientific Explanation* (Cambridge, MA: Harvard University Press, 2008).

11. Michael Strevens, "Précis of *Depth*," *Philosophy and Phenomenological Research* 84 (2012): 447–460, at 455.

12. Ibid., 449–450.

13. Ibid., 455–456.

14. Paul Davies, *The Eerie Silence: Are We Alone in the Universe?* (London: Penguin Kindle Edition, 2010), Kindle Locations 3161–3162.

15. Austin Allen, interview of Paul Davies, http://bigthink.com/ideas/20035.

16. Davies, *The Eerie Silence*, Kindle Locations 3720–3728.

17. Steven Weinberg, *Dreams of a Final Theory: The Scientist's Search for the Ultimate Laws of Nature* (New York: Vintage, 1994).

18. For a philosophical introduction to quantum gravity and its problems, see Steven Weinstein and Dean Rickles, "Quantum Gravity," in *The Stanford Encyclopedia of Philosophy* (spring 2011 ed.), ed. Edward N. Zalta, http://plato.stanford.edu/archives/spr2011/entries/quantum-gravity/.

19. For an accessible discussion of string theory and the Theory of Everything, see Brian Greene, *The Elegant Universe: Superstrings, Hidden Dimensions, and the Quest for the Ultimate Theory* (New York: Vintage, 2000).

20. See, e.g., Lee Smolin, *The Trouble with Physics* (Boston: Houghton Mifflin Harcourt, 2006).

21. David Deutsch, *The Beginning of Infinity: Explanations That Transform the World* (London: Allen Lane Science, 2011).

22. See Richard Dawkins's TED talk, "Why the Universe Seems So Strange," http://www.ted.com/talks/richard_dawkins_on_our_queer_universe.html.

23. Deutsch, *The Beginning of Infinity*, 53.

24. Ibid., 456.

6   The Moral Case against Radical Life Extension

1. See http://supercentenarian.com/oldest/jeanne-calment.html.

2. Nicholas Agar, *Humanity's End: Why We Should Reject Radical Enhancement* (Cambridge, MA: MIT Press, 2010), chapters 5–6.

3. For a very useful survey of SENS, see Aubrey de Grey and Michael Rae, *Ending Aging: The Rejuvenation Breakthroughs That Could Reverse Human Aging in Our Lifetime* (New York: St. Martin's Press, 2007). See also the SENS foundation website at http://www.sens.org/.

4. Aubrey de Grey, "An Engineer's Approach to the Development of Real Anti-aging Medicine," http://www.sens.org/files/pdf/manu16.pdf.

5. This is essentially the conclusion of Siddhartha Mukherjee's *The Emperor of All Maladies: A Biography of Cancer* (New York: Scribner, 2010). At the conclusion of his account, Mukherjee chooses to see cancer as a "normal" part of human biology. He says, "We are inherently destined to slouch toward a malignant end."

6. See Jason Pontin, "Is Defeating Aging Only a Dream? No One Has Won Our $20,000 Challenge to Disprove Aubrey de Grey's Anti-aging Proposals," *Technology Review* 109 (2006): 80–84, http://www.technologyreview.com/sens/index.aspx.

7. Daniel Callahan, *What Price Better Health? Hazards of the Research Imperative* (Berkeley: University of California Press, 2003).

8. Ibid., 74.

9. Ibid.

10. See Christopher Boorse, "On the Distinction between Disease and Illness," *Philosophy and Public Affairs* 5 (1975): 49–68.

11. See Wayne Miller, *King of Hearts: The True Story of the Maverick Who Pioneered the Open Heart Surgery* (New York: Times Books, 2000), for a fascinating but harrowing account of the beginnings of open-heart surgery.

12. Ronald Bailey, "Is Longevity Research Inherently Immoral?," http://reason.com/blog/2012/01/27/is-longevity-research-inherently-immoral.

13. Ibid. See also http://www.tasciences.com/.

14. Allen Buchanan, *Better Than Human: The Promise and Perils of Enhancing Ourselves* (Oxford: Oxford University Press, 2011), 129.

15. Ibid.

16. See http://www.tasciences.com/.

17. See http://www.tasciences.com/patients/.

18. Ibid.

7   A Defense of Truly Human Enhancement

1. George Annas, "The Man on the Moon," in *Science Fiction and Philosophy: From Time Travel to Superintelligence*, ed. Susan Schneider (Malden, MA: Wiley Blackwell, 2009), 227–240.

2. Allen Buchanan, Dan Brock, Norman Daniels, and Daniel Wikler, *From Chance to Choice: Genetics and Justice* (Cambridge: Cambridge University Press, 2000), 152.

3. See John Harris, *Wonderwoman and Superman: The Ethics of Human Biotechnology* (Oxford: Oxford University Press, 1992), for a comparison of educational and genetic enhancement.

4. Francis Galton, *Inquiries into Human Faculty and Its Development* (London: Macmillan, 1883), 17, n. 1.

5. For excellent histories of eugenics, see Daniel Kevles, *In the Name of Eugenics: Genetics and the Uses of Human Heredity* (Cambridge, MA: Harvard University Press, 1998), and Diane Paul, *Controlling Human Heredity: 1865 to the Present* (Atlantic Highlands, NJ: Humanity Books, 1995).

6. Philip Kitcher, *The Lives to Come: The Genetic Revolution and Human Possibilities* (London: Penguin, 1997).

7. K. Anders Ericsson, R. Krampe, and C. Tesch-Romer, "The Role of Deliberate Practice in the Acquisition of Expert Performance," *Psychological Review* 100 (1993): 363–406.

8. John Horner and James Gorman, *How to Build a Dinosaur: Extinction Doesn't Have to Be Forever* (New York: Dutton, 2009).

9. According to the 10,000 hour rule popularized by Malcolm Gladwell in his book *Outliers: The Story of Success* (New York: Little, Brown, 2008), 10,000 hours of deliberate practice should guarantee mastery over almost any human activity.

10. Jürgen Habermas, *The Future of Human Nature* (Cambridge: Polity, 2003), 51.

11. Ibid., 48.

8 Why Radical Cognitive Enhancement Will (Probably) Enhance Moral Status

1. Allen Buchanan, "Moral Status and Human Enhancement," *Philosophy and Public Affairs 37* (2009): 346–381.

2. Mark Walker, "Enhancing Genetic Virtue: A Project for Twenty-First Century Humanity?" *Politics and Life Sciences* 28 (2009): 27–47.

3. Julian Savulescu and Ingmar Persson, "The Perils of Cognitive Enhancement and the Urgent Imperative to Enhance the Moral Character of Humanity," *Journal of Applied Philosophy 25* (2008): 162–177; Julian Savulescu and Ingmar Persson, *Unfit for the Future? Modern Technology, Liberal Democracy, and the Need for Moral Enhancement* (Oxford: Oxford University Press, 2012).

4. Thomas Douglas, "Moral Enhancement," *Journal of Applied Philosophy* 25 (2008): 228–245.

5. Buchanan, "Moral Status and Human Enhancement," 346.

6. Some sources for Buchanan's views about personhood and moral status are Stephen Darwall, *The Second Person Standpoint: Morality, Respect, and Accountability* (Cambridge, MA: Harvard University Press, 2006), and T. M. Scanlon, *What We Owe to Each Other* (Cambridge, MA: Belknap Press of Harvard University Press, 1998).

7. Buchanan, "Moral Status and Human Enhancement," 347.

8. Ibid., 359.

9. See, e.g., David DeGrazia, "Genetic Enhancement, Post-Persons, and Moral Status: A Reply to Buchanan," *Journal of Medical Ethics* 38 (2012): 135–139.

10. Buchanan, "Moral Status and Human Enhancement," 357.

11. Thomas Douglas, "Cognitive Enhancement and the Supra-Personal Moral Status," *Philosophical Studies* 162 (2013): 473–497, offers a very useful taxonomy of possible patterns that status enhancements may follow. The weak thresholds described here correspond to Douglas's Plateau model.

12. Jeff McMahan, "Cognitive Disability and Cognitive Enhancement," *Metaphilosophy* 40 (2009): 582–605.

13. Buchanan, "Moral Status and Human Enhancement," 359.

14. Ibid., 363.

15. Toward the end of DeGrazia's discussion, he presents an alternative analysis of stories like his "Future with Post-Persons." See David DeGrazia, "Genetic Enhancement, Post-Persons, and Moral Status: A Reply to Buchanan," *Journal of Medical Ethics* 38 (2012): 135–139 at 138–139. This alternative account replaces talks of differing moral statuses with talk of differing moral interests. The analysis in terms of differing moral interests is deserving of moral discussion. However, since the bulk of DeGrazia's paper talks explicitly of moral statuses I do not address it here.

16. Ibid., 137.

17. Ibid., 135.

18. Ibid., 137.

19. Nicholas Agar, "Why We Can't Really Say What Post-Persons Are," *Journal of Medical Ethics* 38 (2012): 144–145; David DeGrazia, "Genetic Enhancement, Post-Persons, and Moral Status: Author Reply to Commentaries," *Journal of Medical Ethics* 38 (2012): 145–147.

20. DeGrazia, "Genetic Enhancement, Post-Persons, and Moral Status: Author Reply to Commentaries," 146.

21. Judith Thomson, "A Defense of Abortion," *Philosophy and Public Affairs* 1 (1971): 47–66.

22. See Ian Parker, Annals of Philanthropy, "The Gift," *New Yorker*, August 2, 2004, 54.

23. McMahan, "Cognitive Disability and Cognitive Enhancement," 603.

24. Ibid.

25. Ibid., 603–604.

26. Ibid., 604.

27. See Susan Dominus, "Could Conjoined Twins Share a Mind?," *New York Times*, May 25, 2011, http://www.nytimes.com/2011/05/29/magazine/could-conjoined-twins-share-a-mind.html.

28. Douglas, "Human Enhancement and Supra-Personal Moral Status," 482.

29. Ibid., 482–483.

30. Michael Smith, *The Moral Problem* (Oxford: Blackwell, 1994).

31. For an entertaining account of SETI, see Paul Davies, *The Eerie Silence: Are We Alone in the Universe?* (London: Penguin Kindle Edition, 2010).

9   Why Moral Status Enhancement Is a Morally Bad Thing

1. Francis Fukuyama, *Our Posthuman Future: Consequences of the Biotechnology Revolution* (New York: Farrar, Straus & Giroux, 2002).

2. Julian Savulescu and Ingmar Persson, *Unfit for the Future? Modern Technology, Liberal Democracy, and the Need for Moral Enhancement* (Oxford: Oxford University Press, 2012).

3. Jeff McMahan, "Cognitive Disability and Cognitive Enhancement," *Metaphilosophy* 40 (2009): 582–605, at 598–601.

4. See, e.g., Peter Singer *Practical Ethics*, 2nd ed. (Cambridge: Cambridge University Press, 1993).

5. Thomas Douglas, "The Harms of Status Enhancement Could Be Compensated or Outweighed: A Response to Agar," forthcoming in *Journal of Medical Ethics*.

6. Ray Kurzweil, *The Singularity Is Near: When Humans Transcend Biology* (London: Penguin, 2005).

7. Thomas Douglas, "Cognitive Enhancement and the Supra-Personal Moral Status," *Philosophical Studies* 162 (2013): 473–497.

8. Ibid., 488.

9. Julian Savulescu and Ingmar Persson, *Unfit for the Future? Modern Technology, Liberal Democracy, and the Need for Moral Enhancement* (Oxford: Oxford University Press, 2012).

10. Douglas, "Cognitive Enhancement and the Supra-Personal Moral Status," 488.

11. Ibid., 489.

10   A Technological Yet Truly Human Future—as Depicted in *Star Trek*

1. James Hughes, *Citizen Cyborg: Why Democratic Societies Must Respond to the Redesigned Human of the Future* (Cambridge, MA: Westview, 2004).

2. For widely read examples of prominent scientists' engagement with the issue of the first human, see Richard Leakey's book written with Roger Lewin, *Origins Reconsidered: In Search of What Makes Us Human* (New York: Doubleday, 1992), and Donald Johanson, *Lucy: The Beginnings of Humankind* (New York: Simon & Schuster, 1981).

Index

Basic Bioethics
Arthur Caplan, editor

Books Acquired under the Editorship of Glenn McGee and Arthur Caplan

Peter A. Ubel, *Pricing Life: Why It's Time for Health Care Rationing*

Mark G. Kuczewski and Ronald Polansky, eds., *Bioethics: Ancient Themes in Contemporary Issues*

Suzanne Holland, Karen Lebacqz, and Laurie Zoloth, eds., *The Human Embryonic Stem Cell Debate: Science, Ethics, and Public Policy*

Gita Sen, Asha George, and Piroska Östlin, eds., *Engendering International Health: The Challenge of Equity*

Carolyn McLeod, *Self-Trust and Reproductive Autonomy*

Lenny Moss, *What Genes Can't Do*

Jonathan D. Moreno, ed., *In the Wake of Terror: Medicine and Morality in a Time of Crisis*

Glenn McGee, ed., *Pragmatic Bioethics, 2nd edition*

Timothy F. Murphy, *Case Studies in Biomedical Research Ethics*

Mark A. Rothstein, ed., *Genetics and Life Insurance: Medical Underwriting and Social Policy*

Kenneth A. Richman, *Ethics and the Metaphysics of Medicine: Reflections on Health and Beneficence*

David Lazer, ed., *DNA and the Criminal Justice System: The Technology of Justice*

Harold W. Baillie and Timothy K. Casey, eds., *Is Human Nature Obsolete? Genetics, Bioengineering, and the Future of the Human Condition*

Robert H. Blank and Janna C. Merrick, eds., *End-of-Life Decision Making: A Cross-National Study*

Norman L. Cantor, *Making Medical Decisions for the Profoundly Mentally Disabled*

Margrit Shildrick and Roxanne Mykitiuk, eds., *Ethics of the Body: Post-Conventional Challenges*

Alfred I. Tauber, *Patient Autonomy and the Ethics of Responsibility*

David H. Brendel, *Healing Psychiatry: Bridging the Science/Humanism Divide*

Jonathan Baron, *Against Bioethics*

Michael L. Gross, *Bioethics and Armed Conflict: Moral Dilemmas of Medicine and War*

Karen F. Greif and Jon F. Merz, *Current Controversies in the Biological Sciences: Case Studies of Policy Challenges from New Technologies*

Deborah Blizzard, *Looking Within: A Sociocultural Examination of Fetoscopy*

Ronald Cole-Turner, ed., *Design and Destiny: Jewish and Christian Perspectives on Human Germline Modification*

Holly Fernandez Lynch, *Conflicts of Conscience in Health Care: An Institutional Compromise*

Mark A. Bedau and Emily C. Parke, eds., *The Ethics of Protocells: Moral and Social Implications of Creating Life in the Laboratory*

Jonathan D. Moreno and Sam Berger, eds., *Progress in Bioethics: Science, Policy, and Politics*

Eric Racine, *Pragmatic Neuroethics: Improving Understanding and Treatment of the Mind-Brain*

Martha J. Farah, ed., *Neuroethics: An Introduction with Readings*

Jeremy R. Garrett, ed., *The Ethics of Animal Research: Exploring the Controversy*

Books Acquired under the Editorship of Arthur Caplan

Sheila Jasanoff, ed., *Reframing Rights: Bioconstitutionalism in the Genetic Age*

Christine Overall, *Why Have Children? The Ethical Debate*

Yechiel Michael Barilan, *Human Dignity, Human Rights, and Responsibility: The New Language of Global Bioethics and Bio-Law*

Tom Koch, *Thieves of Virtue: When Bioethics Stole Medicine*

Timothy F. Murphy, *Ethics, Sexual Orientation, and Choices about Children*

Daniel Callahan, *In Search of the Good: A Life in Bioethics*

Robert Blank, *Intervention in the Brain: Politics, Policy, and Ethics*

Gregory E. Kaebnick and Thomas H. Murray, eds., *Synthetic Biology and Morality: Artificial Life and the Bounds of Nature*

Dominic A. Sisti, Arthur L. Caplan, and Hila Rimon-Greenspan, eds., *Applied Ethics in Mental Healthcare: An Interdisciplinary Reader*

Barbara K. Redman, *Research Misconduct Policy in Biomedicine: Beyond the Bad-Apple Approach*

Russell Blackford, *Humanity Enhanced: Genetic Choice and the Challenge for Liberal Democracies*

Nicholas Agar, *Truly Human Enhancement: A Philosophical Defense of Limits*